Min Ru

Minimal Surfaces through Nevanlinna Theory

De Gruyter Studies in Mathematics

Volume 92

Min Ru

Minimal Surfaces through Nevanlinna Theory

DE GRUYTER

Mathematics Subject Classification 2020
32H30, 52A10

Author
Prof. Min Ru
University of Houston
Department of Mathematics
Houston, TX 77204
USA
mru2@Central.UH.EDU

ISBN 978-3-11-099982-2
e-ISBN (PDF) 978-3-11-098955-7
e-ISBN (EPUB) 978-3-11-098996-0
ISSN 0179-0986

Library of Congress Control Number: 2023931747

Bibliographic information published by the Deutsche Nationalbibliothek
The Deutsche Nationalbibliothek lists this publication in the Deutsche Nationalbibliografie; detailed bibliographic data are available on the Internet at http://dnb.dnb.de.

Typesetting: VTeX UAB, Lithuania
Printing and binding: CPI books GmbH, Leck

www.degruyter.com

Dedicated to my lovely granddaughter Riley Ru

Preface

In the mid-1960s, Chern and Osserman [20] began a program to study the value distribution properties of the Gauss maps of minimal surfaces immersed in $\mathbb{R}^m$ through the theory of holomorphic curves (Nevanlinna theory). This program proved crucial when Fujimoto [27], solved the long-standing "five-point theorem" Conjecture in 1988, which states that *the classical Gauss map of a nonflat complete minimal surface immersed in* $\mathbb{R}^3$ *can omit at most four distinct values*. Before that, Xavier [88] already obtained a weaker "seven-point theorem" by using the logarithmic derivative lemma in the Nevanlinna theory. Since then, there have been many important developments in this area and this book describes the progress of this program.

The theory of minimal surfaces is naturally linked to the complex analysis. Under isothermal coordinates, a surface is minimal if and only if its parameterizations are harmonic. By considering the minimal surface as a complex Riemann surface, the Weierstrass–Enneper representation associates it with a pair of functions f, g, where f is analytic and g is meromorphic. The function g is globally defined, and by regarding g as the map $g : M \to \mathbb{P}^1(\mathbb{C})$, it becomes the Gauss map of the minimal surface M after we identify the unit sphere S^2 with $\mathbb{P}^1(\mathbb{C})$ using the stereographic projection. In this way, we can apply the theory of complex analysis, particularly the Nevanlinna theory, to study minimal surfaces.

This book is intended to be self-contained, with few prerequisites making it suitable for undergraduate students as well as graduate students and researchers. Chapters 1 and 2 introduce the mathematics necessary to investigate minimal surfaces, including the basics of differential geometry and complex analysis. Chapter 2 also includes the theory of algebraic curves, serving as an introduction to the theory of holomorphic curves (Nevanlinna theory) and playing a crucial rule in the study of Gauss maps of the complete minimal surfaces with finite total curvature. Chapter 3 focuses the heart of the book, which is the the value distribution of the Gauss maps of minimal surfaces immersed in $\mathbb{R}^3$. Chapter 4 gives a detailed presentation of the Nevanlinna theory and Chapter 5 studies the Gauss maps of minimal surfaces immersed $\mathbb{R}^m$. Finally in Chapter 6, we establish the nonintegrated defect relation for the meromorphic maps of complete Kähler manifolds whose universal cover is the ball $B(R_0)$ $(0 < R_0 \leq \infty)$ into $\mathbb{P}^N(\mathbb{C})$ following Xaiver's method developed in his proof of the "seven-point theorem".

I would like to acknowledge the financial support of the U. S. National Science Foundation (through grants DMS-9596181, DMS-9800361) and U.S. National Security Agency (through grants MDA904-99-1-0034, MDA904-01-1-0051, MDA904-03-1-0067, H98230-05-1-0042, H98230-07-1-0050, H98230-09-1-0004, H98230-11-1-0201), as well as the Simons Foundation Mathematics and Physical Sciences-Collaboration Grants for Mathematicians (award numbers 521604 and 916140). I am grateful to my students Qili Cai and Chin

https://doi.org/10.1515/9783110989557-201

Jui Yang for reading a preliminary version with corrections and suggestions. Lastly, I would like to thank my wife Yu Shen, my son Aaron Ru, and my daughter Christina Ru for their love, support, understanding, and patience. During the time the book was written, my granddaughter Riley Ru was born, and I dedicate this book to her.

Contents

1 Some background in differential geometry

The goal of this chapter is developing some basics of differential geometry needed to investigate minimal surfaces in $\mathbb{R}^3$ and $\mathbb{R}^n$.

1.1 Curves in $\mathbb{R}^n$

We begin with some basic notions. A *parameterized curve* in $\mathbb{R}^n$ is a map $\boldsymbol{\alpha} : I \to \mathbb{R}^n$, where I is an interval in $\mathbb{R}$. It is said to be *regular* if $\boldsymbol{\alpha}'(t) \neq 0$ for all $t \in I$. We can write $\boldsymbol{\alpha}(t) = (x_1(t), x_2(t), \ldots, x_n(t))$. For example, $\boldsymbol{\alpha}(t) = (\cos t, \sin t, t)$ for $t \in \mathbb{R}$ is a parameterization of a helix.

The *tangent vector* (or velocity vector) of $\boldsymbol{\alpha}$ is $\boldsymbol{\alpha}'(t) = (x_1'(t), \ldots, x_n'(t))$. The *arc length* of a curve on $[a, b]$ is

$$\text{length}(\boldsymbol{\alpha}) = \int_a^b \|\boldsymbol{\alpha}'(t)\| dt.$$

A curve $\boldsymbol{\alpha}(t)$ is of *unit speed* if $\|\boldsymbol{\alpha}'(t)\| \equiv 1$. Any nonunit speed regular curve $\boldsymbol{\alpha}(t)$ can be reparameterized by changing the parameter t to the arc-length parameter s, i. e., $s(t) = \int_a^t \|\boldsymbol{\alpha}'(u)\| du$, to form a unit speed curve $\boldsymbol{\beta}(s) := \boldsymbol{\alpha}(t(s))$, where $t = t(s)$ is the inverse of the function $s = s(t)$. Therefore we can always assume that the curves we will be discussing are unit-speed curves. This assumption means that we are only interested in the geometric shape of a regular curve since reparameterizing does not change the shape of a curve.

Let $\boldsymbol{\alpha} : I \to \mathbb{R}^3$ be a unit-speed curve. We are interested in the rate of change of the tangent vector. Denote by $\mathbf{T}(s) := \boldsymbol{\alpha}'(s)$ the unit-tangent vector. Then the *curvature of* $\boldsymbol{\alpha}$ is

$$\kappa(s) := \|\mathbf{T}'(s)\| = \|\boldsymbol{\alpha}''(s)\|.$$

Denote $\mathbf{N}(s) := \frac{\mathbf{T}'(s)}{\|\mathbf{T}'(s)\|}$, the principal normal of $\boldsymbol{\alpha}$, and $\mathbf{B}(s) = \mathbf{T}(s) \times \mathbf{N}(s)$, the binormal of $\boldsymbol{\alpha}$. We have the following Frenet formulas:

$$\mathbf{T}' = \kappa\mathbf{N}, \quad \mathbf{N}' = -\kappa\mathbf{T} + \tau\mathbf{B}, \quad \mathbf{B}' = -\tau\mathbf{N},$$

where $\tau(s)$ given by $\mathbf{B}'(s) = -\tau(s)\mathbf{N}(s)$ is the *torsion* of $\boldsymbol{\alpha}$.

Example. What is the curvature of a circle at point p?

The location of p does not matter because of the symmetry of the circle. A circle of radius r can be parameterized by $\boldsymbol{\alpha}(t) = (r\cos t, r\sin t, 0)$, $0 < t < 2\pi$. Notice that this is not a unit-speed curve since $\boldsymbol{\alpha}'(t) = (-r\sin t, r\cos t, 0)$, and hence

https://doi.org/10.1515/9783110989557-001

$\|\boldsymbol{\alpha}'(t)\| = r$. So we reparameterize it into a unit-speed curve by letting $\frac{s}{r} = t$ to get $\boldsymbol{\beta}(s) = (r\cos(s/r), r\sin(s/r), 0)$, $0 < s < 2\pi r$. Therefore

$$\boldsymbol{\beta}'(s) = (-\sin(s/r), \cos(s/r), 0),$$
$$\boldsymbol{\beta}''(s) = \left(-\frac{1}{r}\cos(s/r), -\frac{1}{r}\sin(s/r), 0\right).$$

Thus

$$\kappa(s) = \|\boldsymbol{\beta}''(s)\| = \frac{1}{r},$$

that is, the curvature of a circle of radius r is $\frac{1}{r}$. Notice that if r is large, then the curvature is small, whereas if r is small, then the curvature is large.

1.2 Surfaces in $\mathbb{R}^3$ and $\mathbb{R}^n$

A (local) *parameterized surface* in $\mathbb{R}^n$ is a one-to-one map $\mathbf{x} : U \subset \mathbb{R}^2 \to \mathbb{R}^n$, where U is an open subset of $\mathbb{R}^2$. In this book, we always assume that $\mathbf{x}$ is of class at least C^2, that is, the partial derivatives of $\mathbf{x}$ up to the second order all exist and are continuous. We can write $\mathbf{x}(u,v) = (x_1(u,v), x_2(u,v), \ldots, x_n(u,v))$, $(u,v) \in U$. The surface is said to be *regular* if $\mathbf{x}_u \times \mathbf{x}_v \neq \mathbf{0}$ for all points $(u,v) \in U$. A global surface in $\mathbb{R}^n$ is a set M such that for every point $p \in M$, there is a local parameterization $\mathbf{x} : U \subset \mathbb{R}^2 \to M \subset \mathbb{R}^n$ with $p \in \mathbf{x}(U)$.

We now focus on the surfaces M in $\mathbb{R}^3$ and introduce various notions of curvatures for M. A *curve on a surface* M is a mapping from an interval to the surface, that is, $\boldsymbol{\alpha} : I \subset \mathbb{R} \to M \subset \mathbb{R}^3$. We can write $\boldsymbol{\alpha}(t) = \mathbf{x}(u(t), v(t))$ for some functions $u, v : I \to \mathbb{R}$. Given $p \in M$, a *tangent vector* to M at $p \in M$ is the tangent vector of some curve $\boldsymbol{\alpha} : (-\epsilon, \epsilon) \to M$ (for some $\epsilon > 0$) with $\boldsymbol{\alpha}(0) = p$, that is, it is $\boldsymbol{\alpha}'(0)$. The *tangent space* of M at $p \in M$, denoted by T_pM, consists of all of the tangent vectors to M at $p \in M$, that is,

$$T_pM = \{\boldsymbol{\alpha}'(0) \mid \boldsymbol{\alpha} : (-\epsilon, \epsilon) \to M \text{ for some } \epsilon > 0 \text{ with } \boldsymbol{\alpha}(0) = p\},$$

which is a vector space with standard basis $\{\mathbf{x}_u|_p, \mathbf{x}_v|_p\}$. The unit normal to the surface $\mathbf{n}$ is given by

$$(\mathbf{n} \circ \mathbf{x}) = \frac{\mathbf{x}_u \times \mathbf{x}_v}{\|\mathbf{x}_u \times \mathbf{x}_v\|}.$$

The first fundamental form (also called a *metric*) I assigns to every $p \in M$ the map $I_p : T_pM \times T_pM \to \mathbb{R}$ given by $I_p(v,w) = v \cdot w$ for all $v, w \in T_pM$, where $v \cdot w$ is the standard dot product of v, w (as vectors in $\mathbb{R}^3$). If we write $v = a\mathbf{x}_u + b\mathbf{x}_v$ and $w = c\mathbf{x}_u + d\mathbf{x}_v$, then $I(v,w) = v \cdot w = (ac)g_{11} + (ad + bc)g_{12} + (bd)g_{22}$, where $g_{11} = \mathbf{x}_u \cdot \mathbf{x}_u$, $g_{12} = g_{21} = \mathbf{x}_u \cdot \mathbf{x}_v$, $g_{22} = \mathbf{x}_v \cdot \mathbf{x}_v$. For this reason, we call $g_{11}, g_{12}, g_{21}, g_{22}$ the *coefficients of the first fundamental form.*

We now ready to introduce various notions of the curvatures of surfaces. As we have seen above, the definition of the curvature for curves used the (unit) tangent vector of the curves. However, in the surface case, $\dim T_pM = 2$, so the choices of the tangent vectors for M are not canonical. Instead, we use its unit-normal $\mathbf{n}$ to define the curvatures. We use two ways to introduce various notions of curvatures for surfaces: (1) through curves on the surfaces and (2) through the shape operator $S_p := -d\mathbf{n}|_p$. We first discuss the approach of using curves on the given surfaces.

The normal curvature of curves in a surface

Let $\boldsymbol{\alpha}$ be a unit-speed curve on a surface M. We introduce the *Darboux frame* along the curve, $\{\boldsymbol{\alpha}'(t), \mathbf{n}(t), \mathbf{n}(t) \times \boldsymbol{\alpha}'(t)\}$, where $\mathbf{n}(t)$ is the unit normal to the surface restricted to the $\boldsymbol{\alpha}(t)$ (i. e., $\mathbf{n}(t) = \mathbf{n}(\boldsymbol{\alpha}(t))$). It is a moving orthonormal frame, so every other vector can be written as a linear combination of the vectors in the Darboux frame. Notice that $\boldsymbol{\alpha}''$ is perpendicular to $\boldsymbol{\alpha}'$, so we can write

$$\boldsymbol{\alpha}''(t) = \kappa_n(t)\mathbf{n}(t) + \kappa_g(t)(\mathbf{n}(t) \times \boldsymbol{\alpha}'(t)),$$

where $\kappa_n(t) = \boldsymbol{\alpha}''(t) \cdot \mathbf{n}(t)$, $\kappa_g(t) = \boldsymbol{\alpha}''(t) \cdot (\mathbf{n}(t) \times \boldsymbol{\alpha}'(t))$. The scalars $\kappa_n(t)$ and $\kappa_g(t)$ are called the *normal curvature* and the *geodesic curvature* of $\boldsymbol{\alpha}$ at the point $p = \boldsymbol{\alpha}(t)$. We have $\kappa^2(t) = \kappa_n^2(t) + \kappa_g^2(t)$, where $\kappa(t)$ is the curvature of $\boldsymbol{\alpha}$.

The normal curvature of surfaces with respect to a given tangent vector $v \in T_pM$

Let M be a surface in $\mathbb{R}^3$, $p \in M$, and let $v \in T_pM$ be a unit tangent vector (direction). Then there is a (unit-speed) curve $\boldsymbol{\alpha} : (0, \epsilon) \to M$ such that $\boldsymbol{\alpha}(t_0) = p$ and $\boldsymbol{\alpha}'(t_0) = v$ for some $t_0 \in (0, \epsilon)$. We define the *normal curvature of M at p in the direction v*, denoted by $\kappa_{n,v}(p)$, as the normal curvature of the curve $\boldsymbol{\alpha}$ at p, that is,

$$\kappa_{n,v}(p) := \boldsymbol{\alpha}''(t_0) \cdot \mathbf{n}(p),$$

where $\mathbf{n}(p)$ is the unit normal at the point p. The normal curvature depends on the directions. The $\max \kappa_{n,v}(p)$ and $\min \kappa_{n,v}(p)$, where max and min are taken over all $v \in T_pM$ with $\|v\| = 1$, are called the *principal curvatures*. Note that the above definition uses the curves in a surface. Theorem 1.2.1 below tells us that $\kappa_{n,v}(p)$ is also related to the shape operator S_p.

The Gauss map and shape operator

There is a unique, up to a direction, unit normal $\mathbf{n}(p)$ to the surface M at $p \in M$, which is normal to the tangent plane T_pM. We fix a direction of $\mathbf{n}$ (in this case, we say that M is oriented, which we always assume). So $v \in T_pM$ if and only if $v \cdot \mathbf{n}(p) = 0$, that is, v is perpendicular to $\mathbf{n}(p)$. The *Gauss map* of the surface M is the map $\mathbf{n} : M \to S^2 \subset \mathbb{R}^3$ that sends every point $p \in M$ to the unit normal $\mathbf{n}(p)$ to M at the point p. Define the *shape operator*

$$S_p = -d\mathbf{n}|_p.$$

A key observation about S_p is that, in fact, $S_p(v)$ is inside T_pM for every $v \in T_pM$, and it is a symmetric linear operator, that is, $S_p(v) \cdot w = v \cdot S_p(w)$ for all $v, w \in T_pM$. The *second fundamental form of M*, denoted by II, assigns to each $p \in M$ the map $\mathrm{II}_p : T_pM \times T_pM \to \mathbb{R}$ given by $\mathrm{II}_p(v, w) = w \cdot S_p(v)$ for $v, w \in T_pM$. Note that $S_p(\mathbf{x}_u) = -d\mathbf{n}(\mathbf{x}_u) = -\mathbf{n}_u$ and $S_p(\mathbf{x}_v) = -\mathbf{n}_v$, and hence under the standard basis $\{\mathbf{x}_u, \mathbf{x}_v\}$,

$$\mathrm{II}_p(a\mathbf{x}_u + b\mathbf{x}_v, c\mathbf{x}_u + d\mathbf{x}_v) = (ac)h_{11} + h_{12}(bc + ad) + h_{22}(bd),$$

where $h_{11} = \mathrm{II}_p(\mathbf{x}_u, \mathbf{x}_u), h_{12} = h_{21} = \mathrm{II}_p(\mathbf{x}_u, \mathbf{x}_v), h_{22} = \mathrm{II}_p(\mathbf{x}_v, \mathbf{x}_v)$. We call the data $\{h_{11}, h_{12}, h_{21}, h_{22}\}$ the *coefficients of the second fundamental form.* Note that, by the definition,

$$h_{11} = \mathbf{x}_u \cdot S_p(\mathbf{x}_u) = -\mathbf{x}_u \cdot \mathbf{n}_u = \mathbf{n} \cdot \mathbf{x}_{uu},$$

where in the last equation we used the fact that $\mathbf{x}_u \cdot \mathbf{n} = 0$, so $(\mathbf{x}_u \cdot \mathbf{n})_u = 0$, which gives $-\mathbf{x}_u \cdot \mathbf{n}_u = \mathbf{n} \cdot \mathbf{x}_{uu}$. Similarly, $h_{12} = -\mathbf{n}_u \cdot \mathbf{x}_v = \mathbf{n} \cdot \mathbf{x}_{uv}$ and $h_{22} = -\mathbf{x}_v \cdot \mathbf{n}_v = \mathbf{n} \cdot \mathbf{x}_{vv}$.

The following theorem links the shape operator with the normal curvature.

Theorem 1.2.1.

$$\kappa_{n,v}(p) = S_p(v) \cdot v.$$

Proof. Let $\mathbf{x} : U \to M$ be a parameterization of M. Take a (unit-speed) curve $\boldsymbol{\alpha} : (0, \epsilon) \to M$ on M with $\boldsymbol{\alpha}(t_0) = p$ and $\boldsymbol{\alpha}'(t_0) = v$ for $0 < t_0 < \epsilon$. Write $\boldsymbol{\alpha}(t) = \mathbf{x}(u(t), v(t))$. Then $\boldsymbol{\alpha}'(t) = u'(t)\mathbf{x}_u(t) + v'(t)\mathbf{x}_v(t)$, where $\mathbf{x}_u(t) = \mathbf{x}_u(u(t), u(t))$ and $\mathbf{x}_v(t) = \mathbf{x}_v(u(t), u(t))$. Hence

$$\begin{aligned}\boldsymbol{\alpha}''(t) &= u'(t)^2\mathbf{x}_{uu}(t) + 2u'(t)v'(t)\mathbf{x}_{uv}(t) + v'(t)^2\mathbf{x}_{vv}(t)\\ &\quad + u''(t)\mathbf{x}_u(t) + v''(t)\mathbf{x}_v(t).\end{aligned}$$

Therefore

$$\begin{aligned}\kappa_{n,v}(p) &= \boldsymbol{\alpha}''(t_0) \cdot \mathbf{n}(p)\\ &= u'(t_0)^2(\mathbf{x}_{uu}(t_0) \cdot \mathbf{n}(p)) + 2u'(t_0)v'(t_0)(\mathbf{x}_{uv}(t_0) \cdot \mathbf{n}(p))\\ &\quad + v'(t_0)^2(\mathbf{x}_{vv}(t_0) \cdot \mathbf{n}(p))\\ &= h_{11}(p)u'(t_0)^2 + 2h_{12}(p)u'(t_0)v'(t_0) + h_{22}(p)v'(t_0)^2\\ &= \mathrm{II}_p(\boldsymbol{\alpha}'(t_0), \boldsymbol{\alpha}'(t_0)) = \mathrm{II}_p(v, v) = S_p(v) \cdot v.\end{aligned}$$

□

Since S_p is a symmetric linear transformation, from the linear algebra we know that real eigenvalues of S_p always exist. Let κ_1 and κ_2 be eigenvalues of S_p with unit eigenvectors $v_1, v_2 \in T_pM$, that is, $S_p(v_1) = \kappa_1 v_1$ and $S_p(v_2) = \kappa_2 v_2$. Then

$$\kappa_{n,v_1}(p) = S_p(v_1) \cdot v_1 = \kappa_1 \|v_1\|^2 = \kappa_1,$$

and, similarly,

$$\kappa_{n,v_2}(p) = \kappa_2,$$

that is, κ_1 and κ_2 are the (special) normal curvatures of M at $p \in M$ in the directions of v_1 and v_2, respectively. They are indeed (see Euler's formula below) the maximal and minimal curvatures among all normal curvatures of M at p. Therefore the eigenvalues κ_1 and κ_2 of S_p are the *principal curvatures* of M at p. We call the corresponding unit eigenvectors the *principal directions*. Note that if $\kappa_1 \neq \kappa_2$, then v_1 and v_2 are orthogonal.

Theorem 1.2.2 (Euler's formula). *Let* $\mathbf{e}_1$ *and* $\mathbf{e}_2$ *be the unit vectors in the principal directions at p corresponding to the principal curvatures* κ_1 *and* κ_2. *Suppose that* $v = \cos\theta \mathbf{e}_1 + \sin\theta \mathbf{e}_2$ *for some* $\theta \in [0, 2\pi)$. *Then*

$$\kappa_{n,v} = \kappa_1 \cos^2\theta + \kappa_2 \sin^2\theta.$$

Proof. By Theorem 1.2.1 we have

$$\begin{aligned}
\kappa_{n,v} &= S_p(v) \cdot v \\
&= (S_p(\cos\theta \mathbf{e}_1 + \sin\theta \mathbf{e}_2)) \cdot (\cos\theta \mathbf{e}_1 + \sin\theta \mathbf{e}_2) \\
&= \cos^2\theta S_p(\mathbf{e}_1) \cdot \mathbf{e}_1 + 2\cos\theta \sin\theta S_p(\mathbf{e}_1) \cdot \mathbf{e}_2 + \sin^2\theta S_p(\mathbf{e}_2) \cdot \mathbf{e}_2 \\
&= \cos^2\theta \kappa_1 \mathbf{e}_1 \cdot \mathbf{e}_1 + 2\cos\theta \sin\theta \kappa_1 \mathbf{e}_1 \cdot \mathbf{e}_2 + \sin^2\theta \kappa_2 \mathbf{e}_2 \cdot \mathbf{e}_2 \\
&= \kappa_1 \cos^2\theta + \kappa_2 \sin^2\theta,
\end{aligned}$$

where we used the fact that $S_p(\mathbf{e}_i) = \kappa_i \mathbf{e}_i$, $i = 1, 2$, and $\mathbf{e}_1 \cdot \mathbf{e}_2 = 0$ (see the remark above), and $\|\mathbf{e}_i\| = 1$, $i = 1, 2$. This finishes the proof. □

The mean curvature and Gauss curvature for surfaces in $\mathbb{R}^3$

Let M be a surface in $\mathbb{R}^3$, and let $p \in M$. Let κ_1 and κ_2 be the principal curvatures of M at p. We define the *Gaussian curvature* of M at p as

$$K(p) = \kappa_1 \kappa_2$$

and the *mean curvature* of M at p as

$$H(p) = \frac{\kappa_1 + \kappa_2}{2}.$$

Since κ_1 and κ_2 are eigenvalues of S_p, we have

$$K(p) = \kappa_1\kappa_2 = \det(S_p)$$

and

$$H(p) = \frac{\kappa_1 + \kappa_2}{2} = \frac{1}{2}\mathrm{trace}(S_p).$$

We have

$$K = \frac{\det(h_{ij})}{\det(g_{ij})} = \frac{h_{11}h_{22} - h_{12}^2}{g_{11}g_{22} - g_{12}^2}, \quad H = \frac{h_{11}g_{22} - 2h_{12}g_{12} + h_{22}g_{11}}{2(g_{11}g_{22} - g_{12}^2)}. \tag{1.1}$$

The following theorem gives a geometric interpretation of the Gauss curvature K.

Theorem 1.2.3. *Let $p \in M$ with $K(p) \neq 0$. Then*

$$K(p) = \lim_{\Omega \to p} \frac{\text{Area } \mathbf{n}(\Omega)}{\text{Area } \Omega}$$

for any Ω containing and shrinking to p.

Proof. The area of Ω is

$$\iint_{\Omega} \|\mathbf{x}_u \times \mathbf{x}_v\| du dv,$$

and the area of $\mathbf{n}(\Omega)$ is

$$\iint_{\Omega} \|d\mathbf{n}(\mathbf{x}_u) \times d\mathbf{n}(\mathbf{x}_v)\| du dv.$$

It is easy to check that

$$d\mathbf{n}(\mathbf{x}_u) \times d\mathbf{n}(\mathbf{x}_v) = K(\mathbf{x}_u \times \mathbf{x}_v).$$

This gives us the result. □

The Gauss curvature is intrinsic

We prove the Gauss theorem egregium: *The Gauss curvature is intrinsic, that is, it only depends on the metric* (g_{ij}). To do so, let $\mathbf{x}(u^1, u^2)$ be a parameterization of a surface M. We write $\mathbf{x}_1 := \mathbf{x}_{u^1}, \mathbf{x}_2 := \mathbf{x}_{u^2}, \mathbf{x}_{i1} := \frac{\partial \mathbf{x}_i}{\partial u^1}, \mathbf{x}_{i2} := \frac{\partial \mathbf{x}_i}{\partial u^2}$ for $i = 1.2$. Write

$$\mathbf{x}_{ij} = \sum_{k=1}^{2} \Gamma_{ij}^k \mathbf{x}_k + h_{ij}\mathbf{n}, \tag{1.2}$$

where $\Gamma_{ij}^k, 1 \le i, j, k \le 2$, are called the *Christoffel symbols*, and $h_{ij} = \mathbf{n} \cdot \mathbf{x}_{ij}$ are the coefficients of the second fundamental form. Note that the Christoffel symbols are symmetric with respect to the lower indices:

$$\Gamma_{ij}^k = \Gamma_{ji}^k \quad \text{for } 1 \le i, j, k \le 2.$$

Let $g_{ij} := \mathbf{x}_i \cdot \mathbf{x}_j$ be the first fundamental form (also called the metric). We first derive the formula

$$\Gamma_{ij}^k = \frac{1}{2} \sum g^{kt} \left(\frac{\partial g_{it}}{\partial u^j} + \frac{\partial g_{jt}}{\partial u^i} - \frac{\partial g_{ij}}{\partial u^t} \right), \tag{1.3}$$

where (g^{kt}) is the inverse matrix of (g_{ij}). Indeed, from the above expression of $\mathbf{x}_{ij}$ in (1.2), taking the dot product with $\mathbf{x}_l$ (notice that $\mathbf{x}_l \cdot \mathbf{n} = 0$), we get, for $1 \le i, j, l \le 2$,

$$\mathbf{x}_{ij} \cdot \mathbf{x}_l = \sum_{k=1}^{2} \Gamma_{ij}^k g_{kl}.$$

Therefore

$$\frac{\partial g_{it}}{\partial u^j} = \frac{\partial (\mathbf{x}_i \cdot \mathbf{x}_t)}{\partial u^j} = \mathbf{x}_{ij} \cdot \mathbf{x}_t + \mathbf{x}_i \cdot \mathbf{x}_{tj} = \sum_{k=1}^{2} \Gamma_{ij}^k g_{kt} + \sum_{k=1}^{2} \Gamma_{tj}^k g_{ki}.$$

Cyclical permutation of the indices, recalling that $\Gamma_{ij}^k = \Gamma_{ji}^k$, yields

$$\frac{\partial g_{jt}}{\partial u^i} = \sum_{k=1}^{2} \Gamma_{ij}^k g_{kt} + \sum_{k=1}^{2} \Gamma_{ti}^k g_{kj},$$

$$\frac{\partial g_{ij}}{\partial u^t} = \sum_{k=1}^{2} \Gamma_{ti}^k g_{kj} + \sum_{k=1}^{2} \Gamma_{tj}^k g_{ki}.$$

Hence

$$\frac{\partial g_{it}}{\partial u^j} + \frac{\partial g_{jt}}{\partial u^i} - \frac{\partial g_{ij}}{\partial u^t} = 2 \sum_{k=1}^{2} \Gamma_{ij}^k g_{kt}.$$

Taking the inverse matrix of (g_{ij}) proves identity (1.3). From this we see that Γ_{ij}^k only depends on the metric g_{ij}, so it is an intrinsic quantity.

Let

$$\Gamma_{ij,k}^m := \frac{\partial \Gamma_{ij}^m}{\partial u^k}.$$

Write, for $1 \le i, j, k, m \le 2$, the *Riemann curvature tensor of the first kind*

$$R_{ijk}^{m} := \Gamma_{ik,j}^{m} - \Gamma_{jk,i}^{m} + \sum_{l=1}^{2}(\Gamma_{ik}^{l}\Gamma_{jl}^{m} - \Gamma_{jk}^{l}\Gamma_{il}^{m}), \tag{1.4}$$

and define the *curvature tensor of the second kind*

$$R_{ijkl} := \sum_{m=1}^{2} R_{ijk}^{m} g_{ml}.$$

Theorem 1.2.4 (Gauss's theorem egregium).

$$K = \frac{R_{1212}}{\det(g_{ij})}. \tag{1.5}$$

Hence K is intrinsic.

Proof. We use the fact that $(\mathbf{x}_{ij})_k = (\mathbf{x}_{ik})_j$.
Step 1: We first compute $(\mathbf{x}_{ij})_k$. We start from

$$\mathbf{x}_{ij} = \sum_{l=1}^{2} \Gamma_{ij}^{l}\mathbf{x}_l + h_{ij}\mathbf{n}. \tag{1.6}$$

Hence

$$\begin{aligned}(\mathbf{x}_{ij})_k &= \frac{\partial \mathbf{x}_{ij}}{\partial u^k} = \left(\sum_{l=1}^{2}\Gamma_{ij}^{l}\mathbf{x}_l + h_{ij}\mathbf{n}\right)_k \\ &= \sum_{l=1}^{2}\Gamma_{ij,k}^{l}\mathbf{x}_l + \sum_{l=1}^{2}\Gamma_{ij}^{l}\mathbf{x}_{lk} + h_{ij,k}\mathbf{n} + h_{ij}\mathbf{n}_k.\end{aligned} \tag{1.7}$$

On the other hand, it is easy to derive that the matrix of S_p with respect to the standard basis $\{\mathbf{x}_1, \mathbf{x}_2\}$ is $\mathcal{F}_I^{-1}\mathcal{F}_{II}$:

$$S_p(\mathbf{x}_i) = \sum_{j,k=1}^{2} h_{ij}g^{jk}\mathbf{x}_k,$$

where (g^{ik}) is the inverse matrix of (g_{ij}). Since

$$\mathbf{n}_i := \frac{\partial \mathbf{n}}{\partial u^i} = d\mathbf{n}(\mathbf{x}_i) = -S_p(\mathbf{x}_i),$$

we get

$$\mathbf{n}_i = -\sum_{j,k=1}^{2} h_{ij}g^{jk}\mathbf{x}_k. \tag{1.8}$$

Thus by combining (1.6), (1.7), and (1.8) we get

$$(\mathbf{x}_{ij})_k = \sum_{m=1}^{2}\left(\Gamma^m_{ij,k} + \sum_{l=1}^{2}\Gamma^l_{ij}\Gamma^m_{lk} - \sum_{l=1}^{2} h_{ij}h_{kl}g^{lm}\right)\mathbf{x}_m + \left(h_{ij,k} + \sum_{l=1}^{2}\Gamma^l_{ij}h_{lk}\right)\mathbf{n}.$$

If we write $(\mathbf{x}_{ij})_k = \sum_{m=1}^{2} A^m_{ijk}\mathbf{x}_m + B_{ijk}\mathbf{n}$, then

$$A^m_{ijk} = \Gamma^m_{ij,k} + \sum_{l=1}^{2}\Gamma^l_{ij}\Gamma^m_{lk} - \sum_{l=1}^{2} h_{ij}h_{kl}g^{lm}.$$

Step 2: Using the fact that $(\mathbf{x}_{ij})_k = (\mathbf{x}_{kj})_i$, we get $A^m_{ijk} = A^m_{kji}$. This gives

$$\Gamma^m_{ij,k} - \Gamma^m_{kj,i} + \sum_{l=1}^{2}(\Gamma^l_{ij}\Gamma^m_{lk} - \Gamma^l_{kj}\Gamma^m_{li}) = \sum_{l=1}^{2}(h_{ij}h_{kl} - h_{kj}h_{il})g^{lm}.$$

Recalling that $\Gamma^k_{ij} = \Gamma^k_{ij}$, by definition (1.4) the last equation becomes

$$R^m_{ikj} = \sum_{l=1}^{2}(h_{ij}h_{kl} - h_{kj}h_{il})g^{lm}.$$

Thus

$$h_{ij}h_{kl} - h_{kj}h_{il} = \sum_{m=1}^{2} g_{lm}R^m_{ikj} = R_{ikjl}.$$

Therefore

$$K = \frac{h_{11}h_{22} - h_{12}^2}{g_{11}g_{22} - g_{12}^2} = \frac{R_{1212}}{\det(g_{ij})},$$

which finishes the proof. □

See also [79, 12] for alternative proofs.

1.3 The Gauss–Bonnet theorem

The Gauss–Bonnet theorem is a fundamental tool in the study of minimal surfaces, especially in the case where the surface is of finite total curvature. It is also a foundation of the theory of Riemann surfaces and the Nevanlinna theory. In this section, we prove the Gauss–Bonnet theorem.

We first give a new proof of the Gauss theorem egregium using the moving frame method. According to the Gram–Schmidt process in linear algebra, we can always convert $\{\mathbf{x}_u, \mathbf{x}_v\}$ into an orthonormal basis $\{\mathbf{e}_1, \mathbf{e}_2\}$. Indeed, we can take $\mathbf{e}_1 = \frac{\mathbf{x}_u}{\|\mathbf{x}_u\|} = \frac{\mathbf{x}_u}{\sqrt{g_{11}}}$ and

$$\mathbf{e}_2 = \frac{\mathbf{x}_v - (\mathbf{x}_v\cdot\mathbf{e}_1)\mathbf{e}_1}{\|\mathbf{x}_v - (\mathbf{x}_v\cdot\mathbf{e}_1)\mathbf{e}_1\|} = \frac{1}{\sqrt{g_{11}g_{22} - g_{12}^2}}\left(\sqrt{g_{11}}\mathbf{x}_v - \frac{g_{12}}{\sqrt{g_{11}}}\mathbf{x}_u\right).$$

Then $\{\mathbf{e}_1, \mathbf{e}_2, \mathbf{e}_3\}$, where $\mathbf{e}_3 = \mathbf{n}$, is an orthonormal moving frame for $\mathbb{R}^3$. Write

$$\begin{aligned}
\mathbf{e}_{1u} &= b^1_{11}\mathbf{e}_1 + b^2_{11}\mathbf{e}_2 + c_{11}\mathbf{e}_3\\
\mathbf{e}_{1v} &= b^1_{12}\mathbf{e}_1 + b^2_{12}\mathbf{e}_2 + c_{12}\mathbf{e}_3\\
\mathbf{e}_{2u} &= b^1_{21}\mathbf{e}_1 + b^2_{21}\mathbf{e}_2 + c_{21}\mathbf{e}_3\\
\mathbf{e}_{2v} &= b^1_{22}\mathbf{e}_1 + b^2_{22}\mathbf{e}_2 + c_{22}\mathbf{e}_3,
\end{aligned}$$

where $b_{ij}, 1 \le i,j \le 2$, are functions on M. From the first equation above, $b^1_{11} = \mathbf{e}_{1u} \cdot \mathbf{e}_1$. Since $\mathbf{e}_1 \cdot \mathbf{e}_1 = 1$, by taking the partial with respect to u we get $2\mathbf{e}_{1u} \cdot \mathbf{e}_1 = 0$. Hence $\mathbf{e}_{1u} \cdot \mathbf{e}_1 = 0$, i. e. $b_{11} = 0$. In a similar way, we can get $b^1_{12} = b^2_{21} = b^2_{22} = 0$. By differentiating $\mathbf{e}_1 \cdot \mathbf{e}_2 = 0$ we get $\mathbf{e}_{1u} \cdot \mathbf{e}_2 = -\mathbf{e}_1 \cdot \mathbf{e}_{2u}$, that is, $b^2_{11} = -b^1_{21}$; similarly, we have $b^2_{12} = -b^1_{22}$. Thus the above equations become

$$\begin{aligned}
\mathbf{e}_{1u} &= b^2_{11}\mathbf{e}_2 + c_{11}\mathbf{e}_3\\
\mathbf{e}_{1v} &= b^2_{12}\mathbf{e}_2 + c_{12}\mathbf{e}_3\\
\mathbf{e}_{2u} &= -b^2_{11}\mathbf{e}_1 + c_{21}\mathbf{e}_3\\
\mathbf{e}_{2v} &= -b^2_{12}\mathbf{e}_1 + c_{22}\mathbf{e}_3.
\end{aligned} \tag{1.9}$$

We claim that

$$K = \frac{1}{\sqrt{g_{11}g_{22} - g^2_{12}}}(c_{11}c_{22} - c_{12}c_{21}). \tag{1.10}$$

Indeed, from $\mathbf{e}_1 \cdot \mathbf{e}_3 = 0$, differentiating with respect to u on both sides, we get $\mathbf{e}_{1u} \cdot \mathbf{e}_3 + \mathbf{e}_1 \cdot \mathbf{e}_{3u} = 0$, so $c_{11} = \mathbf{e}_{1u} \cdot \mathbf{e}_3 = -\mathbf{e}_1 \cdot \mathbf{e}_{3u}$. Similarly, $c_{12} = -\mathbf{e}_1 \cdot \mathbf{e}_{3v}$, $c_{21} = -\mathbf{e}_2 \cdot \mathbf{e}_{3u}$, and $c_{22} = -\mathbf{e}_2 \cdot \mathbf{e}_{3v}$. Hence, also by noticing that $\mathbf{e}_{3u} \cdot \mathbf{e}_3 = \mathbf{e}_{3v} \cdot \mathbf{e}_3 = 0$,

$$\begin{aligned}
\mathbf{e}_{3u} &= -c_{11}\mathbf{e}_1 - c_{21}\mathbf{e}_2\\
\mathbf{e}_{3v} &= -c_{12}\mathbf{e}_1 - c_{22}\mathbf{e}_2.
\end{aligned}$$

Since $\mathbf{x}_u = \sqrt{g_{11}}\mathbf{e}_1$ and $\sqrt{g_{11}}\mathbf{x}_v = \sqrt{g_{11}g_{22} - g^2_{12}}\mathbf{e}_2 + g_{12}\mathbf{e}_1$, we have

$$\mathbf{e}_{3u} = \mathbf{n}_u = -S_p(\mathbf{x}_u) = -\sqrt{g_{11}}S_p(\mathbf{e}_1)$$

and

$$\sqrt{g_{11}}\mathbf{e}_{3v} = \sqrt{g_{11}}\mathbf{n}_v = -\sqrt{g_{11}}S_p(\mathbf{x}_v) = -\sqrt{g_{11}g_{22} - g^2_{12}}S_p(\mathbf{e}_2) - g_{12}S_p(\mathbf{e}_1).$$

Therefore we get

$$S_p(\mathbf{e}_1) = -\frac{1}{\sqrt{g_{11}}}\mathbf{e}_{3u} = \frac{1}{\sqrt{g_{11}}}(c_{11}\mathbf{e}_1 + c_{21}\mathbf{e}_2),$$

$$\begin{aligned}S_p(\mathbf{e}_2) &= \frac{1}{\sqrt{g_{11}g_{22}-g_{12}^2}}(-\sqrt{g_{11}}\mathbf{e}_{3v} - g_{12}S_p(\mathbf{e}_1))\\ &= \frac{\sqrt{g_{11}}}{\sqrt{g_{11}g_{22}-g_{12}^2}}\left(c_{12}\mathbf{e}_1 + c_{22}\mathbf{e}_2 - \frac{g_{12}}{g_{11}}(c_{11}\mathbf{e}_1 + c_{21}\mathbf{e}_2)\right)\\ &= \frac{\sqrt{g_{11}}}{\sqrt{g_{11}g_{22}-g_{12}^2}}\left(\left(c_{12} - \frac{g_{12}}{g_{11}}c_{11}\right)\mathbf{e}_1 + \left(c_{22} - \frac{g_{12}}{g_{11}}c_{21}\right)\mathbf{e}_2\right).\end{aligned}$$

So the matrix of the shape operator S_p with respect to the basis $\gamma = \{e_1, e_2\}$ is

$$[S_p]_\gamma = \begin{pmatrix} \frac{1}{\sqrt{g_{11}}}c_{11} & \frac{\sqrt{g_{11}}}{\sqrt{g_{11}g_{22}-g_{12}^2}}(c_{12} - \frac{g_{12}}{g_{11}}c_{11}) \\ \frac{1}{\sqrt{g_{11}}}c_{21} & \frac{\sqrt{g_{11}}}{\sqrt{g_{11}g_{22}-g_{12}^2}}(c_{22} - \frac{g_{12}}{g_{11}}c_{21}) \end{pmatrix}.$$

Thus

$$\begin{aligned}K = \det(S_p) &= \frac{1}{\sqrt{g_{11}g_{22}-g_{12}^2}}\left(c_{11}\left(c_{22} - \frac{g_{12}}{g_{11}}c_{21}\right) - c_{21}\left(c_{12} - \frac{g_{12}}{g_{11}}c_{11}\right)\right)\\ &= \frac{1}{\sqrt{g_{11}g_{22}-g_{12}^2}}(c_{11}c_{22} - c_{12}c_{21}).\end{aligned}$$

This proves the claim. Using equations (1.9), it is easy to check that

$$\mathbf{e}_{1u}\cdot\mathbf{e}_{2v} - \mathbf{e}_{2u}\cdot\mathbf{e}_{1v} = c_{11}c_{22} - c_{12}c_{21}.$$

Hence

$$\begin{aligned}(b_{11}^2)_v - (b_{12}^2)_u &= -(\mathbf{e}_{2u}\cdot\mathbf{e}_1)_v + (\mathbf{e}_{2v}\cdot\mathbf{e}_1)_u\\ &= -(\mathbf{e}_2)_{uv}\cdot\mathbf{e}_1 - \mathbf{e}_{2u}\cdot\mathbf{e}_{1v} + (\mathbf{e}_2)_{uv}\cdot\mathbf{e}_1 + \mathbf{e}_{2v}\cdot\mathbf{e}_{1u}\\ &= \mathbf{e}_{1u}\cdot\mathbf{e}_{2v} - \mathbf{e}_{2u}\cdot\mathbf{e}_{1v} = c_{11}c_{22} - c_{12}c_{21}.\end{aligned}$$

Thus we obtain the formula for the Gauss curvature

$$K = \frac{(b_{11}^2)_v - (b_{12}^2)_u}{\sqrt{g_{11}g_{22}-g_{12}^2}}, \tag{1.11}$$

which shows that K is intrinsic.

We now derive the formula for computing the geodesic curvature. Let $\boldsymbol{\alpha}(t)$ be a unit-speed curve in M. Let $\mathbf{e}_i(t) = (\mathbf{e}_i \circ \boldsymbol{\alpha})(t)$ for $i = 1, 2$. Then we can write $\boldsymbol{\alpha}'(t) = \cos\theta(t)\mathbf{e}_1(t) + \sin\theta(t)\mathbf{e}_2(t)$, where $\theta(t)$ is the angle from $\mathbf{e}_1(t)$ to $\boldsymbol{\alpha}'(t)$. Notice that the vector $\mathbf{n}(t)\times\boldsymbol{\alpha}'(t)$ is obtained by rotating the vector $\boldsymbol{\alpha}'(t)$ counterclockwise 90 degrees, and hence

$$\mathbf{n}(t) \times \boldsymbol{\alpha}'(t) = \cos\left(\theta(t) + \frac{\pi}{2}\right)\mathbf{e}_1(t) + \sin\left(\theta(t) + \frac{\pi}{2}\right)\mathbf{e}_2(t)$$
$$= -\sin\theta(t)\mathbf{e}_1(t) + \cos\theta(t)\mathbf{e}_2(t).$$

Also,

$$\boldsymbol{\alpha}''(t) = -\theta'(t)\sin\theta(t)\mathbf{e}_1(t) + \theta'(t)\cos\theta(t)\mathbf{e}_2(t)$$
$$+ \cos\theta(t)\mathbf{e}_1'(t) + \sin\theta(t)\mathbf{e}_2'(t).$$

Thus, by the geodesic curvature formula for $\boldsymbol{\alpha}$,

$$\kappa_g(t) = \boldsymbol{\alpha}''(t) \cdot (\mathbf{n}(t) \times \boldsymbol{\alpha}'(t))$$
$$= (\phi_{12}(t) + \theta'(t))(\cos^2\theta(t) + \sin^2\theta(t)) = \phi_{12}(t) + \theta'(t),$$

where $\phi_{12}(t) = \mathbf{e}_1'(t) \cdot \mathbf{e}_2(t)$. So we derived the *Liouville formula*

$$\kappa_g(t) = \phi_{12}(t) + \theta'(t),$$

where $\theta(t)$ is the angle from $\mathbf{e}_1(t)$ to $\boldsymbol{\alpha}'(t)$.

We now ready to prove the following Gauss–Bonnet formula.

Theorem 1.3.1 (Local Gauss–Bonnet theorem for the regions with smooth boundary). *Let M be a parameterized surface given by $\mathbf{x} : U \to \mathbb{R}^3$. Let $\boldsymbol{\alpha}$ be a unit-speed simple smooth closed curve in M, and let R be the interior region bounded by the curve $\boldsymbol{\alpha}$. Then*

$$\iint_R K dA + \int_0^L \kappa_g ds = 2\pi,$$

where L is the total arc length of $\boldsymbol{\alpha}$, K is the Gauss curvature, and dA is the element of area given by $dA = \sqrt{g_{11}g_{22} - g_{12}^2}dudv$.

Proof. By (1.11), $K = \frac{(b_{11}^2)_v - (b_{12}^2)_u}{\sqrt{g_{11}g_{22} - g_{12}^2}}$. Recall the Green theorem in calculus: for any smooth functions $P(u,v)$, $Q(u,v)$ defined around the region $D \subset \mathbb{R}^2$,

$$\iint_D \left(\frac{\partial Q}{\partial u} - \frac{\partial P}{\partial v}\right) dudv = \int_{\partial D} (Pdu + Qdv).$$

Thus, for $D \subset \mathbb{R}^2$ with $\mathbf{x}(D) = R$,

$$\iint_R K dA = \iint_D K\sqrt{g_{11}g_{22} - g_{12}^2}dudv = \iint_D ((b_{11}^2)_v - (b_{12}^2)_u)dudv$$

$$= -\int_{\partial D} (b_{11}^2 du + b_{12}^2 dv) = -\int_0^L ((\mathbf{e}_{1u} \cdot \mathbf{e}_2)u'(s) + (\mathbf{e}_{1v} \cdot \mathbf{e}_2)v'(s))ds$$

$$= -\int_0^L (\mathbf{e}_1'(s) \cdot \mathbf{e}_2(s))ds = -\int_0^L \phi_{12}(s)ds.$$

On the other hand, by the Liouville formula,

$$\int_0^L \kappa_g ds = \int_0^L (\phi_{12}(s) + \theta'(s))ds = \int_0^L \phi_{12}(s)ds + \theta(L) - \phi(0) = \int_0^L \phi_{12}(s)ds + 2\pi,$$

where we used the fact that ∂R is a smooth closed curve, so $\theta(L) - \phi(0) = 2\pi$. This finishes the proof. □

We now consider the case that ∂R is piecewise smooth but not entirely smooth.

Theorem 1.3.2 (Local Gauss–Bonnet). *Let M be a parameterized surface given by $\mathbf{x} : U \to \mathbb{R}^3$. Suppose that $R \subset M$ is a simply connected region. If the curve ∂R has exterior angles $\epsilon_j, j = 1, \ldots, q$, then*

$$\int_{\partial R} \kappa_g ds + \iint_R K dA + \sum_{j=1}^{q} \epsilon_j = 2\pi.$$

Proof. If ∂R is smooth, then $\triangle\theta = 2\pi$, so

$$2\pi = \triangle\theta = \int_{\partial R} \kappa_g ds + \iint_R K dA.$$

When ∂R has corners, the unit tangent vector turns less by the amount $\sum_{j=1}^{q} \epsilon_j$, and so the result follows. This finishes the proof. □

When $R = T$ is a triangle on M (i. e., a region with boundary consisting of three smooth curves) with interior angels A, B, C, it implies (with $\epsilon_1 = \pi - A$, $\epsilon_2 = \pi - B$, and $\epsilon_3 = \pi - C$) the following:

Theorem 1.3.3 (Gauss formula for an embedded triangle). *Let M be a parameterized surface given by $\mathbf{x} : U \to \mathbb{R}^3$. Let T be an embedded triangle on M with interior angels A, B, C. Then*

$$\iint_T K dA + \int_{\partial T} \kappa_g ds = A + B + C - \pi.$$

We now derive the global version of the Gauss–Bonnet theorem. Consider an oriented surface with piecewise smooth boundary. T. Rado proved in 1925 that any such surface M can be triangulated, that is, we may write $M = \bigcup_{\lambda=1}^{m} \triangle_\lambda$, where

(i) $\triangle_\lambda$ is the image of a triangle under a (oriented-preserving) parameterization;
(ii) $\triangle_\lambda \cap \triangle_\nu$ is either an empty or single vertex, or a single edge;
(iii) when $\triangle_\lambda \cap \triangle_\nu$ consists of a single edge, the orientations of the edge are opposite in $\triangle_\lambda$ and $\triangle_\nu$; and
(iv) at most one edge of $\triangle_\lambda$ is contained in the boundary of M.

Definition 1.3.4. Given a triangulation $\mathcal{T}$ of a surface M with V vertices, E edges, and F faces, we define the *Euler characteristic* $\chi(M, \mathcal{T}) = V - E + F$. It is independent of the choice of the triangulation $\mathcal{T}$, so we just denote it by $\chi(M)$.

We can triangulate a disk as follows:

$V - E + F = 5 - 8 + 4 = 1$ $\quad$ $V - E + F = 9 - 18 + 10 = 1$

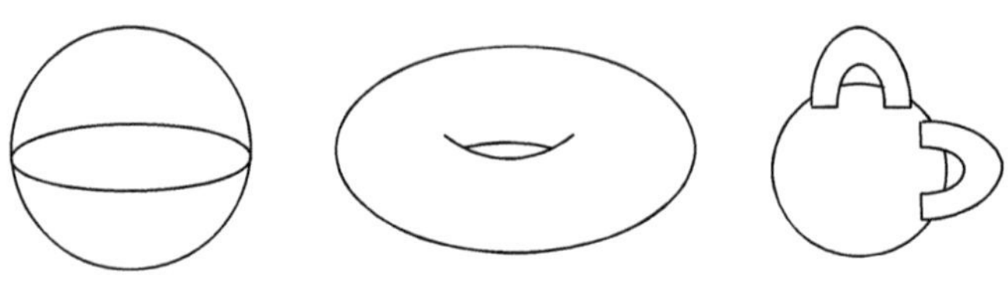

$\chi = 2$ $\quad$ $\chi = 0$ $\quad$ $\chi = -2$

Theorem 1.3.5 (Gauss–Bonnet theorem). *Let M be a compact surface in $\mathbb{R}^3$ without boundary. Then*

$$\iint_M K dA = 2\pi\chi(M),$$

where K is the Gauss curvature, dA is the area measure, and $\chi(M)$ is the Euler characteristic of M.

Proof. Let $M = \bigcup \triangle_\lambda$ be a triangulation. Then

$$\iint_M K dA = \sum_\lambda \iint_{\triangle_\lambda} K dA.$$

Using the local Gauss–Bonnet theorem for triangles $\triangle_\lambda$, we get

$$\iint_{\triangle_\lambda} K dA + \int_{\partial\triangle_\lambda} \kappa_g ds = \sum_{j=1}^{3} \ell_j - \pi,$$

where $\ell_j, 1 \leq j \leq 3$, are the three interior angles of the triangle $\triangle_\lambda$. Since the integrals $\int_{\partial\triangle_\lambda} \kappa_g ds$ cancel in pairs due to the opposite orientations, we have

$$\iint_M K dA = \sum_\lambda \sum_{j=1}^{3} \ell_j - \pi F,$$

where F is the number of triangles $\triangle_\lambda$ (i. e., the number of faces). Notice that at each vertex, the sum of all interior angles is 2π, so

$$\iint_M K dA = 2\pi V - \pi F,$$

where V = # of vertices. Use the fact that M does not have boundary, every triangle has three edges, and each edge is shared by two triangles, and hence $3F = 2E$, so

$$\begin{aligned} \iint_M K dA &= 2\pi V - \pi F = 2\pi V - \pi(2F - 2E) \\ &= 2\pi(E + V - F) = 2\pi\chi(M). \end{aligned}$$

This proves our theorem. □

A typical compact surface is the unit sphere (with no holes), a torus, or a torus with g holes. The number of holes of M, which is also called the *genus* of M, denoted by g, is a topological invariant property (i. e., g does not change if we continuously change M, as long as we do not break it down). We can in fact show that $\chi(M) = 2 - 2g$.

1.4 The abstract surfaces

Recall that a *parameterization* is a one-to-one regular mapping $\mathbf{x} : D \subset \mathbb{R}^2 \to \mathbb{R}^m$ of an open set D. A *surface* in $\mathbb{R}^m$ is a subset $M \subset \mathbb{R}^m$ such that each point of M has a neighborhood (in M) contained in the image of some parameterization

$$\mathbf{x} : D \to M \subset \mathbb{R}^m, \quad \text{where } D \subset \mathbb{R}^2.$$

To define a two-dimensional abstract surface (or a general n-dimensional manifold), we need to remove $\mathbb{R}^m$ in the definition. The following are important ingredients we want to keep: (i) M = set with concept of continuity; (ii) For every $p \in M$, there is subset U of M such that $p \in U$ and we can assign coordinates on U; (iii) Change of coordinates is differentiable (or smooth).

What do we mean by coordinates (in the two-dimensional case)? Coordinate is a way to associate every object (point) we want to study with an ordered tuple of real numbers. For example, under the Cartesian coordinate system, every point on a plane can be associated with a pair of numbers. It is very important, since all mathematics are expressed in numbers.

In our situation the object we want to study is an n-dimensional manifold M, that is, we want to associate every point p with an ordered tuple of real numbers. Also note that for most of times, it suffices to only work on it locally. For example, for the (local) surface $\mathbf{x} : D \subset \mathbb{R}^2 \to M \subset \mathbb{R}^3$, we can associate (locally, i. e., at every point in $\mathbf{x}(D) \subset M$) every point p with $\mathbf{x}^{-1}(p) \in \mathbb{R}^2$. So the inverse $\mathbf{x}^{-1}$ is called the *coordinate map.*

The definition of n-manifolds can be defined in terms of coordinate systems (coordinate maps).

Definition 1.4.1. An n-dimensional differentiable (reps., smooth) manifold is a set M together with a family of one-to-one maps $\phi_\alpha : U_\alpha \subset M \to \mathbb{R}^n$, $\alpha \in \Lambda$, with $\phi_\alpha(U_\alpha)$ being open in $\mathbb{R}^n$ into M, such that
(1) $\bigcup_{\alpha\in\Lambda} U_\alpha = M$, and
(2) For each pair α, β with $U_\alpha \cap U_\beta \neq \emptyset$, the maps

$$\phi_\beta \circ \phi_\alpha^{-1} : \phi_\alpha(U_\alpha \cap U_\beta) \subset \mathbb{R}^n \to \phi_\beta(U_\alpha \cap U_\beta) \subset \mathbb{R}^n$$

and

$$\phi_\alpha \circ \phi_\beta^{-1} : \phi_\beta(U_\alpha \cap U_\beta) \subset \mathbb{R}^n \to \phi_\alpha(U_\alpha \cap U_\beta) \subset \mathbb{R}^n$$

are differentiable (smooth).

The map $\phi_\alpha : U_\alpha \subset M \to \mathbb{R}^n$ is called the *coordinate map*, and its inverse $\phi_\alpha^{-1} : \phi_\alpha(U_\alpha) \to M$ is called a local *parameterization* of M, similarly to our previous discussion in the surface case. An abstract surface is a two-dimensional differentiable manifold.

Example. Consider $S^2 = \{(x, y, z) \mid x^2 + y^2 + z^2 = 1\} \subset \mathbb{R}^3$. Let $N = (0, 0, 1)$ be the north pole, and let $S = (0, 0, -1)$ be the south pole. We consider $\phi_1 : S^2 - \{N\} \to \mathbb{R}^2$ (i. e., $U_1 = S^2 - \{N\}$) be the stereographic projection from the north pole given by

$$\phi_1(x, y, z) = \left(\frac{x}{1-z}, \frac{y}{1-z}\right).$$

Then $\phi_1^{-1} : \mathbb{R}^2 \to S^2$ is given by

$$\phi^{-1}(u, v) = \left(\frac{2u}{1+u^2+v^2}, \frac{2v}{1+u^2+v^2}, \frac{u^2+v^2-1}{1+u^2+v^2}\right).$$

Similarly, let $\phi_2 : S^2 - \{S\} \to \mathbb{R}^2$ (i. e., $U_2 = S^2 - \{S\}$) be the stereographic projection from the south pole given by

$$\phi_2(x,y,z) = \left(\frac{x}{1+z}, \frac{y}{1+z}\right).$$

Then $\phi_1(U_1 \cap U_2) = \mathbb{R}^2 - \{(0,0)\}$, and

$$(\phi_2 \circ \phi_1^{-1})(u,v) = \left(\frac{u}{u^2+v^2}, \frac{v}{u^2+v^2}\right),$$

which is a smooth map. In the same way, we can check that $\phi_1 \circ \phi_2^{-1}$ is also a smooth map. Hence S^2 is a two-dimensional smooth manifold.

Smooth functions and smooth maps

Let M be a smooth differential manifold of dimension m. A functions $f : M \to \mathbb{R}$ is said to be smooth at $p \in M$ if there is a coordinate chart (U,ϕ) such that $f \circ \phi^{-1} : \phi(U) \subset \mathbb{R}^m \to \mathbb{R}$ is smooth (i. e., its partials of any order exist). Similarly, let M and N be smooth differential manifolds of dimensions m and n. A map $f : M \to N$ is said to be smooth at $p \in M$ if there are a coordinate chart (U,ϕ) of p and a coordinate chart (V,ψ) of $q = f(p)$ such that $\psi \circ f \circ \phi^{-1} : \phi(U) \subset \mathbb{R}^m \to \psi(V) \subset \mathbb{R}^n$ is smooth.

Tangent space T_pM

Let M be a smooth differential manifold of dimension m, and let $p \in M$. A smooth curve in M passing through $p \in M$ is a smooth map $\alpha : (-\epsilon,\epsilon) \to M$ such that $\alpha(0) = p$. The key is to define its tangent vector $\alpha'(0)$. Here is the definition: Denote by $C_p^\infty(M)$ the set of real-valued smooth functions defined around $p \in M$; then $\alpha'(0)$ is the map $\alpha'(0) : C_p^\infty(M) \to \mathbb{R}$ such that, for any $f \in C_p^\infty(M)$,

$$\alpha'(0)(f) = \frac{d}{dt}(f \circ \alpha)(t)|_{t=0}.$$

A *tangent vector* v_p to M at $p \in M$ is $v_p = \alpha'(0)$ for some smooth curve $\alpha : (-\epsilon,\epsilon) \to M$ with $\alpha(0) = p$. The tangent space T_pM is the collection of the tangent vectors v_p to M at the point p. We now discuss the standard basis for the tangent space T_pM. Let (U,ϕ) be a local coordinate for M at p. Write $\phi = (x^1,\dots,x^m)$ and the maps $\frac{\partial}{\partial x^i}|_p : C_p^\infty(M) \to \mathbb{R}$ for $1 \le i \le m$ as follows:

$$\left.\frac{\partial}{\partial x^i}\right|_p (f) := \left.\frac{\partial(f \circ \phi^{-1})}{\partial x^i}\right|_{\phi(p)}, \quad f \in C_p^\infty(M);$$

note that $f \circ \phi^{-1}$ is a function defined around $\phi(p) \in \mathbb{R}^m$, and $(x^1,\dots,x^m)$ is the standard Cartesian coordinates of $\mathbb{R}^m$, so the partials $\frac{\partial(f\circ\phi^{-1})}{\partial x^i}|_{\phi(p)}$ are defined for $1 \le i \le m$. Alternatively, let (U,ϕ) be a local coordinate for M at p with $\phi(p) = \mathbf{0}$, and let $\alpha_i : (-\epsilon,\epsilon) \to M$, $1 \le i \le m$, be the curves defined as $\alpha_i(t) = (\phi^{-1} \circ \gamma_i)(t)$, where $\gamma_1 : t \mapsto (t,0,\dots,0), \dots, \gamma_m : t \mapsto (0,0,\dots,t)$ are curves in $\mathbb{R}^m$. Then

$$\left.\frac{\partial}{\partial x^i}\right|_p = \alpha_i'(0),$$

and so $\frac{\partial}{\partial x^i}|_p \in T_pM$. It turns out that they indeed form a basis for T_pM, which is called *the standard basis for T_pM with respect to the coordinate system* $(U;x^i)$.

Metric and Gauss curvature for abstract surfaces

A (Riemannian) *metric* on M is g which assigns, for every point $p \in M$, $g(p) := \langle\cdot,\cdot\rangle_p$, an inner product on T_pM such that $g(p)$ is smooth as p varies. In terms of local coordinates (U,x^i), let $g_{ij} = \langle \partial/\partial x^i, \partial/\partial x^j\rangle$. Then we can write

$$g = \sum g_{ij} dx^i dx^j,$$

where $dx^i dx^j = \frac{1}{2}(dx^i \otimes dx^j + dx^j \otimes dx^i)$ is a symmetric tensor.

Let M be an abstract surface (two-dimensional manifold) with metric g. As above, we define the Christoffel symbols: in terms of local coordinates $(U;x^i)$,

$$\Gamma_{ij}^k = \frac{1}{2}\sum g^{kt}\left(\frac{\partial g_{it}}{\partial x^j} + \frac{\partial g_{jt}}{\partial x^i} - \frac{\partial g_{ij}}{\partial x^t}\right),$$

where (g^{kt}) is the inverse matrix of (g_{ij}). Let

$$\Gamma_{ij,k}^m := \frac{\partial \Gamma_{ij}^m}{\partial x^k}.$$

Write, for $1 \le i,j,k,m \le 2$, the *Riemann curvature tensor of the first kind*

$$R_{ijk}^m := \Gamma_{ik,j}^m - \Gamma_{jk,i}^m + \sum_{l=1}^{2}(\Gamma_{ik}^l \Gamma_{jl}^m - \Gamma_{jk}^l \Gamma_{il}^m),$$

and define the *curvature tensor of the second kind*

$$R_{ijkl} := \sum_{m=1}^{2} R_{ijk}^m g_{ml}.$$

Then the Gauss curvature is defined by

$$K = \frac{R_{1212}}{\det(g_{ij})}. \tag{1.12}$$

S. S. Chern proved that the Gauss–Bonnet theorem still holds for a compact abstract surface without boundary.

2 Analysis and geometry of complex functions

2.1 Holomorphic functions

Denote by $\mathbb{C}$ the set of complex numbers and by $z = x + iy$ complex variables, where $i = \sqrt{-1}$ and $(x, y) \in \mathbb{R}^2$. Let $U \subset \mathbb{C}$ be a domain. Let $f : U \to \mathbb{C}$ be a complex-valued function on U. Let $a \in U$. The function f is said to be *complex differentiable at a* if the complex derivative

$$f'(a) := \lim_{h \to a} \frac{f(a+h) - f(a)}{h}$$

exists. f is said to be *holomorphic on D* if $f'(a)$ exists for every $a \in D$.

Proposition 2.1.1 (The Cauchy–Riemann equations). *Let $f = u + iv$. If f is holomorphic on D, then u and v satisfy the Cauchy–Riemann equations*

$$\frac{\partial u}{\partial x} = \frac{\partial v}{\partial y}, \quad \frac{\partial u}{\partial y} = -\frac{\partial v}{\partial x}.$$

Conversely, suppose that u and v have continuous first-order partial derivatives with respect to x and y on D and u and v satisfy the Cauchy–Riemann equations on D, then f is holomorphic on D.

Proof. "$\Longrightarrow$" Since $f'(z)$ exists, taking the limit as $h \to 0$ with real h, we have

$$f'(z) = \frac{\partial u}{\partial x} + i\frac{\partial v}{\partial x}.$$

Letting $h = ik$ with real k and letting $k \to 0$, we similarly obtain

$$f'(z) = \lim_{k \to 0} \frac{f(z+ik) - f(z)}{ik} = -i\frac{\partial u}{\partial y} + \frac{\partial v}{\partial y}.$$

Equating the two expressions for $f'(z)$, we obtain our result.

"$\Longleftarrow$" Conversely, since the partials of u and v are continuous, by a multivariable calculus theorem, u and v are differentiable at (x, y). This means that

$$u(x+h, y+k) - u(x,y) = \frac{\partial u}{\partial x}h + \frac{\partial u}{\partial y}k + R_u(h,k),$$

and similarly for v, where $R_u(h,k)/(h+ik) \to 0$ as $h + ik \to 0$. Then

$$f(z+h+ik) - f(z) = \left(\frac{\partial u}{\partial x} + i\frac{\partial v}{\partial x}\right)(h+ik) + R_u(h,k) + iR_v(h,k),$$

so

https://doi.org/10.1515/9783110989557-002

$$\lim_{h+ik\to 0}\frac{f(z+h+ik)-f(z)}{h+ik}=\frac{\partial u}{\partial x}+i\frac{\partial v}{\partial x}.$$ □

Define the complex differential operators

$$\frac{\partial}{\partial z}=\frac{1}{2}\left(\frac{\partial}{\partial x}-\sqrt{-1}\frac{\partial}{\partial y}\right),\quad \frac{\partial}{\partial \bar{z}}=\frac{1}{2}\left(\frac{\partial}{\partial x}+\sqrt{-1}\frac{\partial}{\partial y}\right),\quad \partial=\frac{\partial}{\partial z}dz,\quad \bar{\partial}=\frac{\partial}{\partial \bar{z}}d\bar{z},$$

$$d=\partial+\bar{\partial},\quad d^c=\frac{\sqrt{-1}}{4\pi}(\bar{\partial}-\partial),\quad \text{so that } dd^c=\frac{\sqrt{-1}}{2\pi}\partial\bar{\partial}.$$

Corollary 2.1.2. *Let $f=u+iv$. Assume that f is holomorphic on D. Then*

$$\frac{\partial f}{\partial \bar{z}}=0.$$

Furthermore, we have $f'(z)=\frac{\partial f}{\partial z}$. Conversely, if the partials of u and v are continuous and

$$\frac{\partial f}{\partial \bar{z}}=0,$$

then f is holomorphic on D.

We now introduce the Cauchy theorem. Let $\gamma:[a,b]\to\mathbb{C}$ be a piecewise differentiable arc with $\gamma(t)=z(t)$. We define

$$\int_\gamma f(z)dz=\int_a^b f(z(t))z'(t)dt.$$

Note that this integral is invariant under changes of parameter. We will use the following estimate several times:

$$\left|\int_\gamma f(z)dz\right|=\left|\int_a^b f(\gamma(t))\gamma'(t)dt\right|\le\int_a^b|f(\gamma(t))||\gamma'(t)|dt$$
$$\le\left(\sup_{z\in\gamma}|f(z)|\right)\int_a^b|\gamma'(t)|dt.$$

Observe that $\int_a^b|\gamma'(t)|dt$ is the arc length of γ.

Let R be a closed rectangle with boundary ∂R. We say that a function is holomorphic on R if it is holomorphic on a region containing R.

Theorem 2.1.3 (Cauchy's theorem). *If f is holomorphic on R, then*

$$\int_{\partial R} f(z)dz=0.$$

Proof. For any rectangle S, let $\eta(S) = \int_{\partial\eta} f(z)dz$. If we divide R into rectangles R_1, R_2, R_3, R_4, then $\eta(R) = \eta(R_1)+\cdots+\eta(R_4)$, since the integrals over common sides cancel each other. For at least one rectangle R_i, we must have $|\eta(R_i)| \geq \frac{1}{4}|\eta(R)|$. We denote this rectangle S_1 and now divide S_1 into four congruent rectangles and repeat the process. We thus arrive at a sequence $S_0 = R, S_1, S_2, \ldots$ such that $S_{i+1} \subset S_i$ and $|\eta(S_{i+1})| \geq \frac{1}{4}|\eta(S_i)|$, and hence $|\eta(S_n)| \geq 4^{-n}|\eta(R)|$.

Let z^* be the point of intersection of the S_i for $i = 1, 2, \ldots$. Given $\epsilon > 0$, we can choose $\delta > 0$ such that $f(z)$ is analytic in $|z - z^*| < \delta$ and

$$\left|\frac{f(z) - f(z^*)}{z - z^*} - f'(z^*)\right| < \epsilon$$

on $|z - z^*| < \delta$. If n is large enough, then S_n is contained in $|z - z^*| < \delta$. Now noticing that $\int_{S_n} dz = 0$ and $\int_{S_n} zdz = 0$, we have

$$\eta(S_n) = \int_{\partial S_n} (f(z) - f(z^*) - f'(z^*)(z - z^*))dz,$$

and it follows that

$$|\eta(S_n)| \leq \epsilon \int_{\partial S_n} |z - z^*||dz|.$$

In this last integral, $|z - z^*| \leq c_1 2^{-n}$, and $|\int_{\partial S_n} |dz||$, the length of the perimeter of S_n, is $\leq c_2 2^{-n}$, so the last integral is bounded by $c_1 c_2 4^{-n}\epsilon$. This implies that $|\eta(R)| \leq c_1 c_2 \epsilon$; since ϵ is arbitrary, $\eta(R)$ must be 0. □

The following is the Cauchy theorem in a disk. It is sometimes referred to as the Cauchy–Goursat theorem.

Theorem 2.1.4 (Cauchy-Goursat theorem). *Let $D_\rho(z_0)$ be the open disk $|z - z_0| < \rho$. If f is analytic in $D_\rho(z_0)$, then $\int_\gamma f(z)dz = 0$ for every closed curve γ in $D_\rho(z_0)$.*

Now we introduce the Cauchy integral formula. It is easy to verify that if a piecewise differentiable closed curve γ does not pass through the point a, then the value of the integral $\int_\gamma \frac{dz}{z-a}$ is a multiple of $2\pi i$. We define the *index of the point a with respect to γ* or the *winding number of γ with respect to a* by

$$n(\gamma, a) := \frac{1}{2\pi i}\int_\gamma \frac{dz}{z - a}.$$

Theorem 2.1.5 (Cauchy's integral formula). *Let f be a holomorphic function in a domain D, and let γ be a closed curve in D. If $a \notin \gamma$, then*

$$n(\gamma, a)f(a) = \frac{1}{2\pi i}\int_\gamma \frac{f(z)}{z-a}dz.$$

Proof. Let us apply Cauchy's theorem to $F(z) = (f(z) - f(a))/(z - a)$. F is analytic for $z \neq a$. For $z = a$, it satisfies the condition $\lim_{z\to a} F(z)(z - a) = \lim_{z\to a}(f(z) - f(a)) = 0$. Hence

$$\int_\gamma \frac{f(z) - f(a)}{z-a}dz = 0,$$

which is the same as

$$\int_\gamma \frac{f(z)}{z-a}dz = f(a)\int_\gamma \frac{dz}{z-a}.$$

□

Theorem 2.1.6 (Power series expansion). *Let $f : D \to \mathbb{C}$ be holomorphic. Then for each $z_0 \in D$, it can be represented as the power series*

$$f(z) = \sum_{n=0}^{\infty} c_n (z - z_0)^n$$

with nonzero radius of convergence $R > 0$, and

$$c_n = \frac{1}{2\pi i}\int_{\partial B_R(z_0)} \frac{f(w)dw}{(w - z_0)^{n+1}},$$

where $B_R(z_0) \subset D$ is the open disc centered at z_0 with radius R. Moreover, for each specific point $z_0 \in D$, the radius of convergence can be chosen as the minimum distance from point z_0 to the boundary of D.

Remark. For this reason, the holomorphic functions are also called *analytic functions.*

Proof. Assume that f is holomorphic in some ball $B_r(z_0)$. Let $g(z) = f(z + z_0)$. Then g is holomorphic in the ball $B_r(0)$. By Cauchy's formula

$$g(z) = \frac{1}{2\pi i}\int_{\partial B_r(0)} \frac{g(w)dw}{w - z}.$$

We have

$$\frac{1}{w-z} = \frac{1}{w}\frac{1}{1-\frac{z}{w}} = \frac{1}{w}\sum_{n=0}^{\infty}\left(\frac{z}{w}\right)^n,$$

where the last series converges uniformly for all $|z| < r$ since $|z/w| < 1$ by construction. Since the series converges uniformly and all the terms are continuous, we may interchange the order of summation and integration:

$$g(z) = \sum_{n=0}^{\infty} z^n \left(\frac{1}{2\pi i} \int_{\partial B_r(0)} \frac{g(w)dw}{w^{n+1}} \right).$$

Rreturn to the original function f, we conclude that

$$f(z) = g(z - z_0) = \sum_{n=0}^{\infty} c_n (z - z_0)^n,$$

where

$$c_n = \frac{1}{2\pi i} \int_{\partial B_r(z_0)} \frac{f(w)dw}{(w - z_0)^{n+1}}.$$ □

So we see that if f is holomorphic in D, then the derivatives of all orders for f exist, and

$$\begin{aligned} f(z) &= \frac{1}{2\pi i} \int_{\partial D} \frac{f(\zeta)d\zeta}{\zeta - z}, \\ f'(z) &= \frac{1}{2\pi i} \int_{\partial D} \frac{f(\zeta)d\zeta}{(\zeta - z)^2}, \\ f^{(n)}(z) &= \frac{n!}{2\pi i} \int_{\partial D} \frac{f(\zeta)d\zeta}{(\zeta - z)^{n+1}}. \end{aligned} \tag{2.1}$$

Theorem 2.1.7 (Liouville's theorem). *If f is analytic and bounded in $\mathbb{C}$, then f is constant.*

Proof. If $|f|$ is bounded by M, then, by (2.1) with $D(r)$ being the disk of radius r,

$$|f'(z)| \le Mr^{-1}.$$

Letting $r \to \infty$, we see that $f' \equiv 0$. Hence f is constant. □

Assume that f is holomorphic on an open subset of $\mathbb{C}$ with a zero of order m at z_0, that is, $f(z) = (z - z_0)^m h(z)$ locally, and $h(z_0) \neq 0$. Fix some $\rho > 0$ such that $h(z) \neq 0$ in $0 < |z - z_0| \le \rho$. Then, on $0 < |z - z_0| \le \rho$,

$$\frac{f'}{f} = m(z - z_0)^{-1} + \frac{h'}{h}.$$

Thus, by Cauchy's theorem and Cauchy's integral formula,

$$m = \frac{1}{2\pi i} \int_{|z - z_0| = \rho} \frac{f'(z)}{f(z)} dz. \tag{2.2}$$

Theorem 2.1.8 (Hurwitz's theorem). *Let $\{f_k\}$ be a sequence of holomorphic functions on a connected open set $D \subset \mathbb{C}$ that converges uniformly on compact subsets of D to a holomorphic function f that is not constantly zero on D. If f has a zero of order m at z_0, then for every small enough $\rho > 0$ and for sufficiently large $k \in \mathbb{N}$ (depending on ρ), f_k has precisely m zeroes in the disk $|z - z_0| \le \rho$, including multiplicity. Furthermore, these zeroes converge to z_0 as $k \to \infty$.*

Proof. Fix some $\rho > 0$ such that $f(z) \neq 0$ in $0 < |z - z_0| \le \rho$. Choose $\delta > 0$ such that $|f(z)| > \delta$ for z on the circle $|z - z_0| = \rho$. Since $f_k(z)$ converges uniformly on the disc we have chosen, we can find N such that $|f_k(z)| \ge \frac{\delta}{2}$ for every $k \ge N$ and every z on the circle, ensuring that the quotient $f_k'(z)/f_k(z)$ is well defined for all z on the circle $|z - z_0| = \rho$. By (2.1) we can prove that $f_k' \to f'$ uniformly on the disc $|z - z_0| < \rho$, and hence we have another uniform convergence:

$$\frac{f_k'(z)}{f_k(z)} \to \frac{f'(z)}{f(z)}.$$

Denote the number of zeros of $f_k(z)$ in the disk by N_k. By (2.2),

$$m = \frac{1}{2\pi i} \int\limits_{|z-z_0|=\rho} \frac{f'(z)}{f(z)}\, dz = \lim_{k\to\infty} \frac{1}{2\pi i} \int\limits_{|z-z_0|=\rho} \frac{f_k'(z)}{f_k(z)}\, dz = \lim_{k\to\infty} N_k.$$

In the above step, we were able to interchange the integral and limit because of the uniform convergence of the integrand. We have shown that $N_k \to m$ as $k \to \infty$. Since the N_k are integer valued, they must equal m for large enough k. □

2.2 Harmonic functions

A function u is *harmonic* in a region $D \subset \mathbb{R}^2$ if it is of class C^2 (this means that u and its first- and second-order partial derivatives are continuous) and

$$\Delta u = \frac{\partial^2 u}{\partial x^2} + \frac{\partial^2 u}{\partial y^2} = 0.$$

This equation is called the *Laplace equation.*

If u is harmonic, then v is its conjugate harmonic function if $u + iv$ is analytic. In general, there is no single-valued conjugate harmonic function, but there will be in a disk. To show this, let

$$g(z) = \frac{\partial u}{\partial x} - i\frac{\partial u}{\partial y},$$

which is analytic because the Cauchy–Riemann equations are satisfied. Therefore $\int_\gamma g(z)dz = 0$ for every closed curve γ in the disk. Now we can write

$$g(z)dz = \left(\frac{\partial u}{\partial x}dx + \frac{\partial u}{\partial y}dy\right) + i\left(-\frac{\partial u}{\partial y}dx + \frac{\partial u}{\partial x}dy\right).$$

The real part is the differential of u, and the integral over every closed curve of du is 0. Therefore the integral over every closed curve of the imaginary part is 0. Define

$$v(z) = \int_{\gamma}\left(-\frac{\partial u}{\partial y}dx + \frac{\partial u}{\partial x}dy\right),$$

where γ is any curve starting from 0, ending at z, and contained in the unit disk. So v is well defined and has the partials $-\partial/\partial y$ and $\partial u/\partial x$. It is easy to check that v satisfies the Laplace equation and that v is the conjugate harmonic function to u.

Harmonic functions have the mean value property.

Theorem 2.2.1 (Mean value theorem). *If u is harmonic in a disk and $|z - z_0| = r$ is contained in that disk, then*

$$u(z_0) = \frac{1}{2\pi}\int_0^{2\pi} u(z_0 + re^{i\theta})d\theta.$$

Proof. By a change of coordinates we may suppose $z_0 = 0$. Let f be the analytic function which has u as its real part. By Cauchy's integral formula we have

$$f(0) = \frac{1}{2\pi i}\int_{|z|=r}\frac{f(z)}{z}dz = \frac{1}{2\pi i}\int_0^{2\pi}\frac{f(re^{i\theta})}{re^{i\theta}}ire^{i\theta}d\theta.$$

The result follows by taking the real parts. □

2.3 The Ahlfors–Schwarz lemma and the negative curvature method

Theorem 2.3.1 (The maximum principle). *If f is analytic and nonconstant in a region $D \subset \mathbb{C}$, then*

$$|f(z)| < \sup_{\zeta\in D}|f(\zeta)|.$$

Proof. If $\sup_{\zeta\in D}|f(\zeta)| = +\infty$, then the theorem obviously holds. Thus we assume that

$$\sup_{\zeta\in D}|f(\zeta)| = M < +\infty.$$

Assume there exists $z_0 \in D$ such that $|f(z_0)| = M$. If $|f(z)| \equiv M$, then it is easy to prove that f is also constant. Thus there is $z_1 \in D$ such that $|f(z_1)| < M$. Because D is connected, there is a continuous curve $L : z = z(t)$, $t \in [0, 1]$, inside D with $z(0) = z_0$ and $z(1) = z_1$.

Let

$$t^* = \sup\{t \mid |f(z(t))| = M\}.$$

Since f and $z(t)$ are both continuous, we know that $|f(z(t^*))| = M$, $t^* < 1$, and for $t \in [t^*, 1]$,

$$|f(z(t))| < M.$$

Let $z^* = z(t^*) \in D$. Then there exists $\delta > 0$ such that $\overline{B(z^*, \delta)} := \{z \mid |z - z^*| \leq \delta\} \subset D$. By the mean value formula (Cauchy's integral formula),

$$f(z^*) = \frac{1}{2\pi} \int_0^{2\pi} f(z^* + re^{i\theta}) d\theta.$$

Thus

$$M = |f(z^*)| \leq \frac{1}{2\pi} \int_0^{2\pi} |f(z^* + re^{i\theta})| d\theta,$$

that is,

$$\frac{1}{2\pi} \int_0^{2\pi} (|f(z^* + re^{i\theta})| - M) d\theta \geq 0.$$

Because $|f(z^* + re^{i\theta})| - M$ is a nonpositive function, we get $|f(z^* + re^{i\theta})| - M \equiv 0$, that is, $|f(z)| = M$ for all $z \in \overline{B(z^*, \delta)}$. Hence $f(z) = Me^{i\alpha}$ for some real number α. On the other hand, since $z(t)$ is continuous at t^*, $z^* = z(t^*)$, there exists $t_1 \in (t^*, 1)$ such that $z(t_1) \in \overline{B(z^*, \delta)}$, and thus, by the above, $|f(z(t_1))| < M$, a contradiction. Hence

$$|f(z)| < M, \quad z \in D. \qquad \square$$

Theorem 2.3.2 (Schwarz lemma). *Let $D(1)$ be the unit disk on $\mathbb{C}$. Let $f : D(1) \to D(1)$ be a holomorphic map such that $f(0) = 0$. Then $|f(z)| \leq |z|$ for all $z \in D(1)$.*

Proof. Define $g(z) = \frac{f(z)}{z}$ if $z \neq 0$ and $g(0) = f'(0)$. Then g is holomorphic on $D(1)$. Now if $\overline{D(r)} = \{z \mid |z| \leq r\}$ denotes the closed disk of radius r centered at the origin, then the maximum modulus principle implies that for $r < 1$, given any $z \in D(r)$, there exists z_r on the boundary of $\overline{D(r)} = \{z \mid |z| \leq r\}$ such that

$$|g(z)| \leq |g(z_r)| = \frac{|f(z_r)|}{|z_r|} \leq \frac{1}{r}.$$

This proves the theorem by letting $r \to 1$. $\square$

Theorem 2.3.3 (Schwarz–Pick lemma). *Let $f : D(1) \to D(1)$ be a holomorphic map. Then for any two points $z_1, z_2 \in D(1)$, we have*

$$\left|\frac{f(z_1) - f(z_2)}{1 - \overline{f(z_1)}f(z_2)}\right| \leq \left|\frac{z_1 - z_2}{1 - \bar{z}_1 z_2}\right|.$$

In particular, for any $z \in D(1)$,

$$\frac{|f'(z)|}{1 - |f(z)|^2} \leq \frac{1}{1 - |z|^2}.$$

Proof. The proof of the Schwarz–Pick theorem follows from the Schwarz lemma and the fact that a Möbius transformation of the form

$$\frac{z - z_0}{\bar{z}_0 z - 1}, \quad |z_0| < 1,$$

maps the unit circle into itself. Fix z_1 and define the Möbius transformations

$$M(z) = \frac{z_1 - z}{1 - \bar{z}_1 z}, \quad \phi(z) = \frac{f(z_1) - z}{1 - \overline{f(z_1)}z}.$$

Since $M(z_1) = 0$ and the Möbius transformation is invertible, the composition $\phi \circ f \circ M^{-1}$ maps 0 to 0, and the unit disk is mapped into itself. Thus we can apply Theorem 2.3.2 (Schwarz's lemma):

$$\left|(\phi \circ f \circ M^{-1})(z)\right| = \left|\frac{f(z_1) - f(M^{-1}(z))}{1 - \overline{f(z_1)}f(M^{-1}(z))}\right| \leq |z|.$$

Now denoting $z_2 = M^{-1}(z)$ (which will still be in the unit disk) yields the desired conclusion

$$\left|\frac{f(z_1) - f(z_2)}{1 - \overline{f(z_1)}f(z_2)}\right| \leq \left|\frac{z_1 - z_2}{1 - \bar{z}_1 z_2}\right|.$$ □

To reformulate the lemma in terms of geometric quantities, we introduce the following definition.

Definition 2.3.4. Let $\Omega \subset \mathbb{C}$ be a region. A *metric* on Ω is $ds^2 = 2a(z)dzd\bar{z}$ with smooth $a > 0$; $ds^2 = 2b(z)dzd\bar{z}$ is said to be a *pseudo-metric* on Ω if for every $z_0 \in \Omega$, $b(z) = (z - z_0)^k h(z)$ around z_0, where $k \geq 0$, and $h > 0$ is smooth near z_0. The *Gaussian curvature function* $K_{ds^2} : \Omega \to [-\infty, \infty)$ of ds^2 is defined by

$$K_{ds^2}(z) = -\frac{1}{b(z)} \frac{\partial^2 \log b(z)}{\partial z \partial \bar{z}}$$

for $z \in U$-Zero(b). For $z \in$ Zero(b), we define $K_{ds^2}(z) = -\infty$.

Example. Let $D(r)$ be the disc of radius r on $\mathbb{C}$ with center at the origin. The metric

$$ds^2 = \frac{4r^2 dz d\bar{z}}{(r^2 - |z|^2)^2} \tag{2.3}$$

is called the *Poincaré metric* on $D(r)$. We have

$$K \equiv -1.$$

We can reformulate the Schwarz–Pick lemma in terms of the Poincaré metric on the unit disc $D(1)$.

Lemma 2.3.5 (Schwarz–Pick). *Let g_P be the Poincaré metric on the unit disc $D(1)$, and let $f : D(1) \to D(1)$ be a holomorphic map. Then $f^* g_P \le g_P$, that is, the Poincaré distance decreases under any holomorphic map.*

We now prove the more general Ahlfors–Schwarz lemma, which extends the Schwarz–Pick lemma.

Theorem 2.3.6 (Ahlfors–Schwarz lemma). *Let ds^2 denote the Poincaré metric on the unit disc $D(1)$. Let $d\sigma^2$ be a pseudo-metric on $D(1)$ with Gaussian curvature bounded above by -1. Then*

$$d\sigma^2 \le ds^2.$$

Proof. Let $D(r)$ be the disc of radius $r < 1$ with the Poincaré metric ds^2 of curvature -1 given by

$$ds^2 = 2a_r(z) dz d\bar{z}, \quad \text{where } a_r(z) = \frac{2r^2}{(r^2 - |z|^2)^2}.$$

We compare this metric with $d\sigma^2 = 2b(z) dz d\bar{z}$. Put

$$\mu(z) = \log \frac{b(z)}{a_r(z)}.$$

Since $\mu(z) \to -\infty$ as $z \to \partial D(r)$, there is a point $z_0 \in D(r)$ such that

$$\mu(z_0) = \sup\{\mu(z); z \in D(r)\} > -\infty.$$

Then $b(z_0) > 0$. Since z_0 is a maximal point of $\mu(z)$,

$$0 \ge \frac{\partial^2 \mu}{\partial z \partial \bar{z}}(z_0).$$

On the other hand, since the Gaussian curvature of the Poincaré metric is -1 and the curvature of $d\sigma^2$ is bounded above by -1,

$$\frac{\partial^2 \log a_r}{\partial z \partial \bar{z}} = a_r(z) \quad \text{and} \quad \frac{\partial^2 \log b}{\partial z \partial \bar{z}}(z) \geq b(z).$$

So

$$0 \geq \frac{\partial^2 \mu}{\partial z \partial \bar{z}}(z_0) = \frac{\partial^2 \log b}{\partial z \partial \bar{z}}(z_0) - \frac{\partial^2 \log a_r}{\partial z \partial \bar{z}}(z_0) \geq b(z_0) - a_r(z_0).$$

Hence $a_r(z_0) \geq b(z_0)$, and so $\mu(z_0) \leq 0$. By the choice of z_0 we have $\mu(z) \leq 0$ on $D(r)$, that is,

$$a_r(z) \geq b(z).$$

The theorem is proved by letting $r \to 1$. □

To prove the little Picard theorem, we introduce the chordal distance on $\mathbb{C} \cup \{\infty\}$ by

$$|w_1, w_2| = \frac{|w_1 - w_2|}{\sqrt{1 + |w_1|^2}\sqrt{1 + |w_2|^2}}$$

if $w_1, w_2 \neq \infty$ and

$$|w_1, w_2| = \frac{1}{\sqrt{1 + |w_1|^2}}$$

if $w_1 \neq \infty$ and $w_2 = \infty$.

Lemma 2.3.7. *Let g be a meromorphic function on $D(R) = \{|z| < R\}$ $(0 < R \leq +\infty)$ (i. e., g is holomorphic on $D(R)$ except some poles). Let $\alpha \in \mathbb{C}$. Then, for a fixed $\epsilon > 0$, there exists $\delta_0 > 0$ such that for all $\delta \geq \delta_0$, on $D(R) \backslash \{z \mid g(z) = \alpha\}$,*

$$\frac{\partial^2}{\partial z \partial \bar{z}} \log \frac{1}{\log(\delta/|g, \alpha|^2)} \geq \frac{|g'|^2}{(1 + |g|^2)^2}\left(\frac{1}{|g, \alpha|^2 \log^2(\delta/|g, \alpha|^2)} - \epsilon\right)$$

Proof. Let

$$\phi := |g, \alpha|^2 = \frac{|g - \alpha|^2}{(1 + |g|^2)(1 + |\alpha|^2)}.$$

Then we have

$$\frac{\partial \phi}{\partial z} = \frac{g'(\bar{g} - \bar{\alpha})(1 + \alpha \bar{g})}{(1 + |g|^2)^2(1 + |\alpha|^2)},$$

and so

$$\left|\frac{\partial \phi}{\partial z}\right|^2 = \frac{|g'|^2}{(1 + |g|^2)^4} \frac{|g - \alpha|^2}{(1 + |\alpha|^2)^2} |1 + \bar{\alpha} g|^2.$$

Since

$$1-\phi = \frac{|1+\bar{\alpha}g|^2}{(1+|g|^2)(1+|\alpha|^2)},$$

we obtain

$$\left|\frac{\partial\phi}{\partial z}\right|^2 = (\phi-\phi^2)\frac{|g'|^2}{(1+|g|^2)^2}.$$

On the other hand,

$$\frac{\partial^2}{\partial z\partial\bar{z}}\log(1+|g|^2) = \frac{|g'|^2}{(1+|g|^2)^2}.$$

Thus

$$\begin{aligned}\frac{\partial^2}{\partial z\partial\bar{z}}\log\frac{1}{\log(\delta/\phi)} &= \frac{\partial}{\partial z}\left(\frac{1}{\log(\delta/\phi)}\frac{\partial}{\partial\bar{z}}\log\phi\right)\\ &= \frac{-1}{\log(\delta/\phi)}\frac{\partial^2}{\partial z\partial\bar{z}}\log(1+|g|^2) + \frac{1}{\phi^2\log^2(\delta/\phi)}\left|\frac{\partial\phi}{\partial z}\right|^2\\ &= \frac{|g'|^2}{(1+|g|^2)^2}\left(\frac{\phi-\phi^2}{\phi^2\log^2(\delta/\phi)} - \frac{1}{\log(\delta/\phi)}\right)\\ &= \frac{|g'|^2}{(1+|g|^2)^2}\left(\frac{1}{\phi\log^2(\delta/\phi)} - \left(\frac{1}{\log^2(\delta/\phi)} + \frac{1}{\log(\delta/\phi)}\right)\right).\end{aligned}$$

We can take $\delta = \delta_0$ such that $\log^{-2}(\delta/\phi) + \log^{-1}(\delta/\phi) < \epsilon$ because $\phi \le 1$. Thus we get the desired inequality. □

Theorem 2.3.8. *Let g be a nonzero meromorphic function on $D(R) = \{|z| < R\}$. Assume that its image omits $a_1, \ldots, a_q \in \mathbb{C} \cup \{\infty\}$. If $q > 2$, then there exist constants $\delta > 0$ and $C > 0$ such that*

$$\frac{|g'|}{(1+|g|^2)}\prod_{j=1}^{q}\frac{1}{|g,a_j|\log(\delta/|g,a_j|^2)} \le C\frac{R}{R^2-|z|^2}.$$

Proof. After a change of g by a suitable Möbius transformation, we may assume that $a_q = \infty$. Let

$$v := \frac{|g'|}{(1+|g|^2)}\prod_{j=1}^{q}\frac{1}{|g,a_j|\log(\delta/|g,a_j|^2)}.$$

Then

$$v^2 = \frac{(1+|g|^2)^{q-1}w}{\prod_{j=1}^{q}\log^2(\delta/|g,a_j|^2)},$$

where

$$w = \frac{|g'|^2\prod_{j=1}^{q-1}(1+|a_j|^2)}{\prod_{j=1}^{q-1}|g-a_j|^2}.$$

Then for $\epsilon = (q-1)/q$, take δ such that the above lemma holds for all a_j. Thus, also using

$$\frac{\partial^2}{\partial z\partial\bar{z}}\log(1+|g|^2) = \frac{|g'|^2}{(1+|g|^2)^2}$$

and noticing that $\log w$ is harmonic, we have

$$\begin{aligned}\frac{\partial^2}{\partial z\partial\bar{z}}\log v^2 &\geq \frac{|g'|^2}{(1+|g|^2)^2}\left(q-1+\sum_{j=1}^{q}\left(\frac{2}{|g,a_j|^2\log^2(\delta/|g,a_j|^2)}-\frac{q-1}{q}\right)\right)\\ &= \frac{2|g'|^2}{(1+|g|^2)^2}\sum_{j=1}^{q}\frac{1}{|g,a_j|^2\log^2(\delta/|g,a_j|^2)}.\end{aligned}$$

For every fixed $z \in D_R$, it is easy to see that there is at most one j (denoted by j_0) such that $|g(z),a_j| \geq \min_{k<l}|a_k,a_l|/2$, and therefore there exists $C > 0$ such that

$$\begin{aligned}\frac{\partial^2}{\partial z\partial\bar{z}}\log v &\geq \frac{|g'|^2}{(1+|g|^2)^2}\frac{1}{|g,a_{j_0}|^2\log^2(\delta/|g,a_j|^2)}\\ &\geq C\frac{|g'|^2}{(1+|g|^2)^2}\prod_{j=1}^{q}\frac{1}{|g,a_j|^2\log^2(\delta/|g,a_j|^2)} = Cv^2.\end{aligned}$$

The theorem thus follows from the Ahlfors–Schwarz lemma. □

Corollary 2.3.9. *Let g be a nonzero meromorphic function on $D(R) = \{|z| < R\}$. Assume that its image omits $a_1,\ldots,a_q \in \mathbb{C}\cup\{\infty\}$ with $a_q = \infty$. If $q > 2$, then there exists a constant $C > 0$ such that for $0 < \eta < 1$,*

$$\frac{|g'|(1+|g|^2)^{\frac{q-2-q\eta}{2}}}{\prod_{j=1}^{q-1}|g-a_j|^{1-\eta}} \leq C\frac{R}{R^2-|z|^2}. \tag{2.4}$$

Proof. Because the function $\log(\delta x^2)/x^\eta$ $(1 \leq x < +\infty)$ is bounded, we can replace $\log^2(\delta/|g,a_j|^2)$ by $|g,a_j|^{2\eta}$ and thus obtain that

$$\frac{|g'|}{(1+|g|^2)}\frac{1}{\prod_{j=1}^{q}|g,a_j|^{1-\eta}} \leq C\frac{R}{R^2-|z|^2}.$$

Using the fact $a_q = \infty$ and using the definition of the chordal distance that $|w_1, w_2| = \frac{|w_1 - w_2|}{\sqrt{1+|w_1|^2}\sqrt{1+|w_2|^2}}$ if $w_1, w_2 \neq \infty$ and $|w_1, w_2| = \frac{1}{\sqrt{1+|w_1|^2}}$ if $w_1 \neq \infty$ and $w_2 = \infty$, we have the identity

$$\frac{|g'|}{(1+|g|^2)} \frac{1}{\prod_{j=1}^{q} |g, a_j|^{1-\eta}} = C' \frac{|g'|(1+|g|^2)^{\frac{q-2-q\eta}{2}}}{\prod_{j=1}^{q-1} |g - a_j|^{1-\eta}}$$

with $C' = \prod_{j=1}^{q-1}(1+|a_j|^2)^{\frac{1}{2}(1-\eta)}$. □

Theorem 2.3.10 (Little Picard theorem). *Let f be a meromorphic function on $\mathbb{C}$. If the image of f omits at least three points in $\mathbb{C} \cup \{\infty\}$, then f is constant.*

Proof. From (2.4), letting $R \to \infty$, we obtain that $f' \equiv 0$. Thus f is constant. □

In the rest of the section, we introduce the results that will be used in the Chapter 5. The results can also be viewed as the *dual form of the Ahlfors–Schwarz–Pick lemma.* Also see [64, 65].

Lemma 2.3.11. *Let $D(r)$ be the disk of radius r, $0 < r < 1$, and let R be the hyperbolic radius of $D(r)$ in the unit disc (with respect to the Poincaré metric). Let $ds^2 = \lambda(z)^2|dz|^2$ be any conformal metric on $D(r)$ such that the geodesic distance from $z = 0$ to $|z| = r$ is greater than or equal to R. If the Gauss curvature K of the metric ds^2 satisfies*

$$-1 \le K \le 0,$$

then the distance of any point to the origin in the metric ds^2 is greater than or equal to the hyperbolic distance.

Note that the hyperbolic metric in the unit disk is given by

$$d\hat{s}^2 = \hat{\lambda}(z)^2|dz|^2, \quad \hat{\lambda}(z) = \frac{2}{1-|z|^2},$$

and has the curvature $\hat{K} \equiv -1$. The relation between the quantities R and r is therefore given by

$$R = \int_0^r \hat{\lambda}(z)|dz| = \int_0^r \frac{2}{1-t^2}dt = \log\frac{1+r}{1-r},$$

and from Lemma 2.3.11 it follows that

$$\rho(z) \ge \hat{\rho}(z) = \log\frac{1+|z|}{1-|z|},$$

where ρ and $\hat{\rho}$ represent the distances from the point z to the origin in the metric ds^2 and the hyperbolic metric, respectively.

Proof of Lemma 2.3.11. Note first that in the relation above between R and r, we have

$$\frac{dR}{dr} = \frac{2}{1 - r^2} > 0,$$

and we may solve for r in terms of R:

$$r = \frac{e^R - 1}{e^R + 1}, \tag{2.5}$$

or, in general,

$$|z| = \frac{e^{\hat{\rho}(z)} - 1}{e^{\hat{\rho}(z)} + 1}, \tag{2.6}$$

where the right-hand side is increasing in $\hat{\rho}(z)$. We may apply the comparison theorem of Greene and Wu ([38], Prop. 2.15, p. 26) to the metric ds^2 and the hyperbolic metric $d\hat{s}^2$ on the disk $|z| \le r$. The comparison theorem states that for any smooth monotone increasing function f, we have

$$\triangle(f \circ \rho) \le \hat{\triangle}(f \circ \hat{\rho}),$$

where ρ and $\hat{\rho}$ are the distances to the origin in the metrics ds^2 and $d\hat{s}^2$, respectively, $\triangle$ and $\hat{\triangle}$ are the Laplacians with respect to the two metrics, and the two sides are evaluated at points of the same level sets of the two metrics, that is, $\rho = c$ on the left, and $\hat{\rho} = c$ on the right, provided in two dimensions that the Gauss curvatures K and $\hat{K}$ satisfy $0 \ge K \ge \hat{K}$, with a similar condition on the Ricci curvature in higher dimension. In our case, we have $0 \ge K \ge -1 = \hat{K}$, and so we may apply the theorem. Note that the function

$$\log |z| = \log \frac{e^{\hat{\rho}(z)} - 1}{e^{\hat{\rho}(z)} + 1}$$

is harmonic with respect to z and is therefore also harmonic with respect to any conformal metric on $0 < |z| < 1$. In other words, setting

$$f(t) = \log \frac{e^t - 1}{e^t + 1},$$

we have

$$\hat{\triangle}(f \circ \hat{\rho}) \equiv 0$$

for $0 < |z| < 1$. Since f is increasing, we may apply the Greene–Wu comparison theorem to conclude that

$$\triangle(f \circ \rho) \le 0$$

for $0 < |z| < r$, that is, $f \circ \rho$ is superharmonic. For z near 0, we have $\rho(z) \sim \lambda(0)|z|$, and we may apply the minimum principle to the function

$$\log \frac{1}{|z|} f(\rho(z)) = \log \frac{1}{|z|} \frac{e^{\rho(z)} - 1}{e^{\rho(z)} + 1},$$

which is superharmonic in $0 < |z| < r$ and bounded near the origin, to conclude that it takes on its minimum on the boundary $|z| = r$. Since $\rho(z) \geq R$ on $|z| = r$, by (2.5) we have that

$$\log \frac{1}{|z|} \frac{e^{\rho(z)} - 1}{e^{\rho(z)} + 1} \geq \log \frac{1}{r} \frac{e^R - 1}{e^R + 1} = 0$$

for $|z| < r$. Hence

$$\frac{e^{\rho(z)} - 1}{e^{\rho(z)} + 1} \geq |z| = \frac{e^{\hat{\rho}(z)} - 1}{e^{\hat{\rho}(z)} + 1}$$

by (2.6), which implies that $\rho(z) \geq \hat{\rho}(z)$, proving the lemma. □

Lemma 2.3.11 also implies another dual form of the Ahlfors–Schwarz–Pick lemma, closer to the original one.

Lemma 2.3.12. *Let S be a simply connected surface with a complete metric ds^2 whose Gauss curvature satisfies $-1 \leq K \leq 0$. If S is mapped conformally onto the unit disc, then the distance between any two points of S is greater than or equal to the hyperbolic distance between the corresponding points in the disk.*

Proof. Given two points p and q of S, we may map p onto the origin, and let z be the image of the point q. Then the distance between p and q on S is given by $\rho(z)$ in terms of the pullback of the metric on S onto the disk. For any r such that $|z| < r < 1$, let $\hat{\rho}(z)$ be the hyperbolic distance from 0 to z, and let $\rho_r(w)$ be the pullback of the metric on S to $|w| < r$ under the map $z = w/r$. Then since S is complete, we may apply Lemma 2.3.11 to conclude that

$$\hat{\rho}(z) \leq \rho_r(w) = \rho_r(rz).$$

Since $\rho_r(rz) \to \rho(z)$ as $r \to 1$, the lemma is proved. □

Note that Lemma 2.3.12 combined with the standard Ahlfors–Schwarz–Pick lemma implies a generalization of Ahlfors' lemma due to Yau [91] (see also Troyanov [85]): *Let S_1 be a simply connected Riemann surface with a complete metric ds^2 whose Gauss curvature satisfies $-1 \leq K \leq 0$, and let S_2 be a Riemann surface with Gauss curvature bounded above by -1. Let $f : S_1 \to S_2$ be a holomorphic map. Then f is distance decreasing.*

2.4 The Riemann surfaces

A *Riemann surface* M is a one-dimensional complex manifold, that is, it is covered by a family of coordinates $\{U_\alpha, \phi_\alpha\}$ that covers the totality of M, where $\bigcup_\alpha U_\alpha = M$, and $\phi_\alpha : U_\alpha \to \mathbb{C}$ are functions with $\phi_\alpha \circ \phi_\beta^{-1}$ holomorphic on $\phi_\beta(U_\alpha \cap U_\beta)$ (assuming that $U_\alpha \cap U_\beta \neq \emptyset$). Let M be a Riemann surface with a local coordinate $z = x + \sqrt{-1}y$. Denote

$$\frac{\partial}{\partial z} = \frac{1}{2}\left(\frac{\partial}{\partial x} - \sqrt{-1}\frac{\partial}{\partial y}\right), \quad \frac{\partial}{\partial \bar{z}} = \frac{1}{2}\left(\frac{\partial}{\partial x} + \sqrt{-1}\frac{\partial}{\partial y}\right), \quad \partial = \frac{\partial}{\partial z}dz, \quad \bar{\partial} = \frac{\partial}{\partial \bar{z}}d\bar{z},$$

$$d = \partial + \bar{\partial}, \quad d^c = \frac{\sqrt{-1}}{4\pi}(\bar{\partial} - \partial) \quad \text{so that } dd^c = \frac{\sqrt{-1}}{2\pi}\partial\bar{\partial}.$$

The *conformal metric* on M is given by $ds^2 = 2a(dx^2 + dy^2) = 2adzd\bar{z}$ locally, where a is a positive smooth function. Its *Gaussian curvature* is, according to (1.5),

$$K = -\frac{1}{4a} \triangle \log a, \tag{2.7}$$

where $\triangle = \frac{\partial^2}{\partial x^2} + \frac{\partial^2}{\partial y^2} = 4\frac{\partial^2}{\partial z \partial \bar{z}}$ is the usual Laplacian. Let

$$\omega = a(z)\frac{\sqrt{-1}}{2\pi}dz \wedge d\bar{z}$$

be its *associated Kähler form* (or *metric form*). With ω we associate the *Ricci form* $\mathrm{Ric}(\omega) := dd^c \log a$. Then we have

$$\mathrm{Ric}(\omega) = dd^c \log a = -K\omega. \tag{2.8}$$

Assume that M is compact (and without boundary). By the Gauss–Bonnet theorem in Chapter 1 we have

$$-\int_M \mathrm{Ric}(\omega) = \int_M K\omega = \chi(M) = 2 - 2g, \tag{2.9}$$

where g is the genus of M. In the study of holomorphic map $f : M \to N$ (for example, in deriving the Riemann–Hurwitz formula below), where M and N are compact Riemann surfaces, we take a smooth positive $(1,1)$-form ω on N and pull it back to M through f to get a *pseudo-positive* $(1,1)$*-form* $\phi = f^*\omega$, that is, we can write it locally as $\phi = h\frac{\sqrt{-1}}{2\pi}dz\wedge d\bar{z}$ on M with $h(z) = |z|^{2\nu}h_0(z)$ with smooth $h_0 > 0$. In this case, we say that *ϕ has a zero of order ν at $z = 0$*. We have the following Gauss–Bonnet theorem for pseudo-metric, extending (2.9), by slightly modifying the proof (see [76, Theorem A1.1.1]).

Theorem 2.4.1 (Gauss–Bonnet for pseudo-metric). *Let M be a compact Riemann surface. Let $\phi = h\frac{\sqrt{-1}}{2\pi}dz \wedge d\bar{z}$ be a pseudo-positive $(1,1)$-form on M. Let $a_1, \dots, a_k$ be the zeros of ϕ on M with vanishing order $\nu_i, 1 \le i \le k$ (see above for the definition). Then*

$$-\int_M \mathrm{Ric}(\phi) = 2 - 2g + \sum_{j=1}^{k} \nu_j.$$

If we rewrite (in the sense of currents)

$$\int_M \mathrm{Ric}[\phi] = \int_M dd^c[\log h] := \int_M dd^c \log h + \sum_{j=1}^{k} \nu_j,$$

then the theorem becomes

$$\int_M \mathrm{Ric}[\phi] = 2g - 2. \tag{2.10}$$

We now turn to holomorphic maps between two Riemann surfaces. Let $f : M \to N$ be a holomorphic map, where M and N are compact Riemann surfaces. We call $v_f(p)$ the *multiplicity of f at $p \in M$* if there are local coordinates z for M at $p \in M$ with $z(p) = 0$ and w for N at $f(p)$, respectively, with $w(f(p)) = 0$ such that $w = z^{v_f(p)}$.

Theorem 2.4.2 (Riemann–Hurwitz). *Let M and N be compact Riemann surfaces, and let $f : M \to N$ be a holomorphic map. Then*

$$(2g_M - 2) = \deg(f)(2g_N - 2) + \sum_{p\in M} (v_f(p) - 1),$$

where g_M and g_N are the genera of M and N, respectively.

Note that here $\deg(f)$ is the number of $f^{-1}\{a\}$ for some (hence for all) $a \in N$, counting the multiplicities. In other words, if we denote by $|f^{-1}\{a\}|$ the cardinality of the set $f^{-1}\{a\}$, then

$$\deg(f) = |f^{-1}\{a\}| + \sum_{p\in f^{-1}\{a\}} (v_f(p) - 1). \tag{2.11}$$

In the case $N = \mathbb{P}^1(\mathbb{C})$, let ω_{FS} be the Fubini–Study form on $\mathbb{P}^1(\mathbb{C})$. Then $\omega_{\mathrm{FS}} = \frac{1}{(1+|w|^2)^2}\frac{\sqrt{-1}}{2\pi} dw \wedge d\bar{w} = dd^c \log(1 + |w|^2)$ for the affine coordinate $(w, 1) \in \mathbb{P}^1(\mathbb{C})$. Write $f = f_0/f_1$. Then $f^*\omega_{\mathrm{FS}} = dd^c \log(1 + |f_0/f_1|^2)$. Let $p_l, 1 \le l \le k$, be the (distinct) zeros of f_1 on M, and choose holomorphic local coordinates z_l with $z_l(p_l) = 0$. We consider a sufficiently small positive ϵ such that $\overline{U}_l(\epsilon) := \{z_l \mid |z_l| \le \epsilon\}$ are mutually disjoint. Since $dd^c \log |f_1|^2 = 0$ outside $\{p_1, \dots, p_k\}$, we obtain

$$\begin{aligned}\int_M f^*\omega_{\mathrm{FS}} &= \lim_{\epsilon\to 0} \int_{M-\bigcup_{l=1}^k U_l(\epsilon)} f^*\omega_{\mathrm{FS}} \\ &= \lim_{\epsilon\to 0} \int_{M-\bigcup_{l=1}^k U_l(\epsilon)} dd^c \log(1 + |f_0/f_1|^2) = -\lim_{\epsilon\to 0} \sum_{l=1}^{k} \int_{\partial U_l(\epsilon)} d^c \log(1 + |f_0/f_1|^2)\end{aligned}$$

$$= -\lim_{\epsilon\to 0}\sum_{l=1}^{k}\int_{\partial U_l(\epsilon)} d^c \log(|f_0|^2+|f_1|^2) + \lim_{\epsilon\to 0}\sum_{l=1}^{k}\int_{\partial U_l(\epsilon)} d^c \log|f_1|^2$$

$$= \lim_{\epsilon\to 0}\sum_{l=1}^{k}\int_{\partial U_l(\epsilon)} d^c \log|f_1|^2.$$

On the other hand,

$$\sum_{l=1}^{k}\int_{\partial U_l(\epsilon)} d^c \log|f_1|^2 = \sum_{l=1}^{k}\frac{1}{2\pi i}\int_{|z_l|=\epsilon}\frac{f_1'(z)}{f_1(z)}dz = \sum_{l=1}^{k}\nu_{f_1}(p_l) = \deg(f).$$

Hence

$$\deg(f) = \int_M f^*\omega_{\mathrm{FS}}. \tag{2.12}$$

Proof of Theorem 2.4.2. We only prove the case where $N = \mathbb{P}^1(\mathbb{C})$. Let ω_{FS} be the Fubini–Study form on $\mathbb{P}^1(\mathbb{C})$. Then $f^*\omega_{\mathrm{FS}} = h\frac{\sqrt{-1}}{2\pi}dz\wedge d\bar{z}$ is a pseudo-positive (1,1)-form on M whose singular divisor is the ramification divisor (denoted by $\mathrm{ram}(f)$). By (2.10),

$$\int_M dd^c[\log h] = 2g_M - 2,$$

that is,

$$\int_M dd^c \log h + \deg(\mathrm{ram}(f)) = 2g_M - 2.$$

Since by definition

$$\omega_{\mathrm{FS}} = \frac{1}{(1+|w|^2)^2}\frac{\sqrt{-1}}{2\pi}dw\wedge d\bar{w} = dd^c\log(1+|w|^2)$$

for the affine coordinate $(w,1)\in\mathbb{P}^1(\mathbb{C})$, we have $\mathrm{Ric}(\omega_{\mathrm{FS}}) = -2\omega_{\mathrm{FS}}$. Hence

$$-2\deg(f) = -2\int_M f^*\omega_{\mathrm{FS}} = \int_M f^*\mathrm{Ric}(\omega_{\mathrm{FS}}) = \int_M \mathrm{Ric}(f^*\omega_{\mathrm{FS}})$$

$$= \int_M dd^c\log h = 2g_M - 2 - \deg(\mathrm{ram}(f)) = 2g_M - 2 - \sum_{p\in M}(\upsilon(p)-1). \qquad \square$$

For any subset $E\subset M$, define

$$r(E) := \sum_{p\in E}(\upsilon_f(p)-1).$$

Let $a_1, \dots, a_q \in N$, and let $E = f^{-1}\{a_1, \dots, a_q\}$. Then from (2.11) we have $q\deg(f) = |E| + r(E)$, where $|E|$ is the cardinality of the set E. By applying the Riemann–Hurwitz theorem we have

$$(2g_M - 2) = \deg(f)(2g_N - 2) + q\deg(f) - |E| + r(E^c).$$

So we have derived the following theorem.

Theorem 2.4.3 (Algebraic second main theorem (the Riemann relation)). *Let M and N be compact Riemann surfaces, and let $f : M \to N$ be a nonconstant holomorphic map. Let $a_1, \dots, a_q$ be distinct points in N, and let $E = f^{-1}\{a_1, \dots, a_q\} \subseteq M$. Then*

$$(q + 2g_N - 2)\deg(f) \le |E| + 2(g_M - 1).$$

In particular, when $N = \mathbb{P}^1(\mathbb{C})$, we have

$$(q - 2)\deg(f) \le |E| + 2(g_M - 1).$$

2.5 The theory of algebraic curves

In this section, we present the theory of algebraic curves. We mainly follow the method from Chapter 2 of the book *Principle of algebraic geometry* by Griffiths-Harris (see [39]). The n-dimensional complex projective space is $\mathbb{P}^n(\mathbb{C}) = \mathbb{C}^{n+1} - \{0\}/\sim$, where $(a_0, a_1, \dots, a_n) \sim (b_0, b_1, \dots, b_n)$ if and only if $(a_0, \dots, a_n) = \lambda(b_0, \dots, b_n)$ for some $\lambda \in \mathbb{C}$. We denote by $[a_0 : \cdots : a_n]$ the equivalence class of $(a_0, \dots, a_n)$. By an *algebraic curve* we mean a holomorphic map $f : M \to \mathbb{P}^n(\mathbb{C})$, where M is a compact Riemann surface. In this section, we study algebraic curves. The theory is regarded as a model for the general theory of holomorphic curves (Nevanlinna theory).

Let M be a Riemann surface, and let $f : M \to \mathbb{P}^n(\mathbb{C})$ be a holomorphic map. For every point $p \in M$, clearly, we can lift f locally to $\mathbb{C}^{n+1}$, that is, in a neighborhood of any point $p \in M$, we can find a holomorphic vector-valued function $\mathbf{f} = (f_0, \dots, f_n)$ such that $f = [f_0 : \cdots : f_n]$. Such $\mathbf{f}$ is called a *reduced representation* of f at p if $f_0, \dots, f_n$ have no common zeros.

Let $\mathbf{f} = (f_0, \dots, f_n)$ be a reduced representation of f. Define

$$\mathbf{F}_k = \mathbf{f} \wedge \mathbf{f}' \wedge \cdots \wedge \mathbf{f}^{(k)} : M \to \bigwedge^{k+1} \mathbb{C}^{n+1}.$$

Evidently, $\mathbf{F}_{n+1} \equiv 0$. The map $F_k = \mathbb{P}(\mathbf{F}_k) : M \to \mathbb{P}(\bigwedge^{k+1}\mathbb{C}^{n+1}) \subset \mathbb{P}^{N_k}$, where $N_k = \frac{(n+1)!}{(k+1)!(n-k)!} - 1$ and $\mathbb{P}$ is the natural projection, is called *the kth associated map* of f. Note that F_k is well defined, that is, it is independent of the choice of lifting $\mathbf{f}$ and of the local coordinate z.

Let $\{\mathbf{e}_0 = (1,\dots,0),\dots,\mathbf{e}_n = (0,\dots,1)\}$ be the standard basis of $\mathbb{C}^{n+1}$. Then we can write

$$\begin{aligned}\mathbf{F}_k &= \sum_{0\le j_0<\dots<j_k\le n} \det(f_{j_t}^{(l)})_{0\le t\le k,0\le l\le k}\mathbf{e}_{j_0}\wedge\dots\wedge\mathbf{e}_{j_k}\\ &= \sum_{0\le j_0<\dots<j_k\le n} W(f_{j_0},f_{j_1},\dots,f_{j_k})\mathbf{e}_{j_0}\wedge\dots\wedge\mathbf{e}_{j_k},\end{aligned}$$

where $W(f_{j_0},f_{j_1},\dots,f_{j_k}) := \det(f_{j_t}^{(l)})_{0\le t\le k,0\le l\le k}$ is the Wronskian determinant.

Theorem 2.5.1. *The image $f(M)$ lies in a k-dimensional subspace of $\mathbb{P}^n(\mathbb{C})$ but does not lie in any $(k-1)$-dimensional subspace if and only if for any representative $\mathbf{f}$ of f, $\mathbf{F}_k \not\equiv 0$, but $\mathbf{F}_{k+1} \equiv 0$.*

Proof. The statement of this theorem is equivalent to saying that $f(M)$ lies in a $(k+1)$-dimensional subspace of $\mathbb{C}^{n+1}$ but does not lie in any k-dimensional subspace. Letting this be true, without loss of generality, we can assume that $f(M)$ is contained in the plane $\{w_{k+1} = \dots = w_n = 0\}$, which we identify with the space $\mathbb{C}^{k+1}$ of variables $(w_0,\dots,w_k)$. Then all $f_{k+1},\dots,f_n$ are identically zero. It is then clear that $\mathbf{F}_{k+1} \equiv 0$. However, $W(f_0,\dots,f_k) \not\equiv 0$ for, if not, then $f_0,\dots,f_k$ would be linearly dependent, and then its image would lie in some hyperplanes of $\mathbb{C}^{k+1}$, which contradicts the fact that it does not lie in any k-dimensional subspace.

Conversely, assume that $\mathbf{F}_k \not\equiv 0$ but $\mathbf{F}_{k+1} \equiv 0$. Then one coefficient in the expression of $\mathbf{F}_k$ (in terms of $\mathbf{e}_{j_0}\wedge\dots\wedge\mathbf{e}_{j_k}$), say $W(f_0,\dots,f_k) \not\equiv 0$, but $W(f_0,\dots,f_k,f_j) \equiv 0$ for all $j = k+1,\dots,n$. This implies that $f_0,\dots,f_k$ are linearly independent and $f_j = \sum_{t=0}^k b_{jt}f_t$ for $j = k+1,\dots,n$. This means that f is contained in the planes $w_j = \sum_{t=0}^k b_{jt}w_t$, $j = k+1,\dots,n$, so $f(M)$ lies in a $(k+1)$-dimensional subspace of $\mathbb{C}^{n+1}$, and by what we have shown before, it does not lie in any k-dimensional subspace of $\mathbb{C}^{n+1}$. □

From now on, we assume that f is linearly nondegenerate, that is, $f(M)$ is not contained in any hyperplane in $\mathbb{P}^n(\mathbb{C})$. Therefore by Theorem 2.5.1, $\mathbf{F}_k \not\equiv 0$ for $0 \le k \le n$. Let $P \in M$, and let z be a local parameter for M at P with $z(P) = 0$. By making a linear change of coordinates in $\mathbb{C}^{n+1}$ if necessary we can assume that f has a local reduced representation $\mathbf{f} = (f_0,\dots,f_n)$ with

$$f_0 = z^{\delta_0} + \cdots,\dots,f_n = z^{\delta_n} + \cdots, \tag{2.13}$$

such that $0 = \delta_0 < \delta_1 < \dots < \delta_n$. Such an expression is called the *normal form* of f. The integers

$$\nu_i = \delta_i - \delta_{i-1} - 1, \quad 1 \le i \le n, \tag{2.14}$$

are called the *stationary indices of order i* at $z = 0$ (or at the point $P \in M$). Note that

$$\mathbf{f}(z) = (1 + \cdots, z^{1+\nu_1} + \cdots, z^{2+\nu_1+\nu_2} + \cdots, \ldots, z^{n+\nu_1+\cdots+\nu_n} + \cdots).$$

We also have

$$\sum_{i=0}^{n-1}(n-i)\nu_{i+1} + \frac{1}{2}n(n+1) = \delta_1 + \cdots + \delta_n. \tag{2.15}$$

The number $\nu_1(P)$ is the *ramification index of f at P*, which is the order of vanishing of the Jacobian $(\partial f_1/\partial z, \ldots, \partial f_n/\partial z)$ at $z = 0$, that is,

$$\nu_1(P) = \min_{1\le i\le n}(\mathrm{ord}_{z=0}(\partial f_i/\partial z)).$$

Alternatively, $\nu_1(P)$ is a unique integer such that

$$f^*\omega_{\mathrm{FS}} = |z|^{2\nu_1(P)}\lambda_0(z)\frac{\sqrt{-1}}{2\pi}dz \wedge d\bar{z}$$

with some $\lambda_0 > 0$. Similarly, $\nu_{k+1}(P)$ is the ramification index of F_k at P. It is easy to check that

$$F_k{}^*\omega_k = |z|^{2\nu_{k+1}(P)}\lambda_1(z)\frac{\sqrt{-1}}{2\pi}dz \wedge d\bar{z} \tag{2.16}$$

with some $\lambda_1 > 0$ locally, where $\omega_k = dd^c \log \|Z\|^2$ (which is called the *Fubini–Study form on* $\mathbb{P}^{N_k}(\mathbb{C})$), $Z = [x_0 : \cdots : x_{N_k}] \in \mathbb{P}^{N_k}(\mathbb{C})$. Let

$$\beta_i = \sum_{P\in M}\nu_i(P).$$

The number β_i is called the *ith total ramification (stationary) index.*

Write $F_k{}^*\omega_k = h_k(z)\frac{\sqrt{-1}}{2\pi}dz \wedge d\bar{z}$.

Lemma 2.5.2.

$$h_k(z) = \frac{\|\mathbf{F}_{k-1}\|^2\|\mathbf{F}_{k+1}\|^2}{\|\mathbf{F}_k\|^4}$$

for $0 \le k \le n$, *and by convention* $\|\mathbf{F}_{-1}\| \equiv 1$.

Proof. Observe that F_k is holomorphic, so $\Omega_k := F_k{}^*\omega_k$ is smooth, and h_k is nonnegative. Since $\omega_k = dd^c \log \|Z\|^2$,

$$\begin{aligned}\Omega_k = F_k^*\omega_k &= dd^c \log \|\mathbf{F}_k\|^2 = \frac{\sqrt{-1}}{2\pi}\partial\bar{\partial}\log(\mathbf{F}_k, \mathbf{F}_k)\\ &= \frac{\sqrt{-1}}{2\pi}\frac{(\mathbf{F}_k', \mathbf{F}_k')(\mathbf{F}_k, \mathbf{F}_k) - (\mathbf{F}_k', \mathbf{F}_k)(\mathbf{F}_k, \mathbf{F}_k')}{\|\mathbf{F}_k\|^4}dz \wedge d\bar{z}.\end{aligned}$$

Since

$$\mathbf{F}_k = \mathbf{f} \wedge \mathbf{f}' \wedge \cdots \wedge \mathbf{f}^{(k)},$$
$$\mathbf{F}'_k = \mathbf{f} \wedge \mathbf{f}' \wedge \cdots \wedge \mathbf{f}^{(k-1)} \wedge \mathbf{f}^{(k+1)}.$$

For any $z \in \mathbb{C}$ with $\mathbf{F}_{k+1}(z) \neq 0$, by Gram–Schmidt there exist numbers A_{ji} for $0 \leq i \leq j \leq n+1$ with $A_{ii} \neq 0$ and an orthonormal basis $\mathbf{e}_0, \ldots, \mathbf{e}_{k+1}$ such that

$$\mathbf{f}^{(j)}(z) = \sum_{i=0}^{j} A_{ji}\mathbf{e}_i \quad \text{for } j = 0, \ldots, k+1.$$

So

$$\mathbf{F}_j = A_{00} \cdots A_{jj}\mathbf{e}_0 \wedge \cdots \wedge \mathbf{e}_j \quad \text{for } j = 0, \ldots, k+1.$$

Thus

$$\|\mathbf{F}_j\| = |A_{00}| \ldots |A_{jj}| \quad \text{for } j = 0, \ldots, k+1.$$

Also,

$$\mathbf{F}'_k = \left(\prod_{q=0}^{k-1} A_{qq}\right)\mathbf{e}_0 \wedge \cdots \wedge \mathbf{e}_{k-1} \wedge (A_{k+1,k}\mathbf{e}_k + A_{k+1,k+1}\mathbf{e}_{k+1}).$$

Since $\mathbf{e}_0, \ldots, \mathbf{e}_{k+1}$ is an orthonormal basis,

$$(\mathbf{F}'_k, \mathbf{F}'_k) = \left(\prod_{q=0}^{k-1} |A_{qq}|^2\right)(|A_{k+1,k}|^2 + |A_{k+1,k+1}|^2),$$
$$(\mathbf{F}_k, \mathbf{F}_k) = |A_{00}|^2 \cdots |A_{kk}|^2,$$

and

$$(\mathbf{F}'_k, \mathbf{F}_k)(\mathbf{F}_k, \mathbf{F}'_k) = \left(\prod_{q=0}^{k-1} |A_{qq}|^4\right)|A_{kk}|^2|A_{k+1,k}|^2.$$

Therefore

$$\begin{aligned}&(\mathbf{F}'_k, \mathbf{F}'_k)(\mathbf{F}_k, \mathbf{F}_k) - (\mathbf{F}'_k, \mathbf{F}_k)(\mathbf{F}_k, \mathbf{F}'_k)\\ &= |A_{00}|^4 \ldots |A_{k-1,k-1}|^4|A_{kk}|^2|A_{k+1,k+1}|^2 = \|\mathbf{F}_{k-1}\|^2\|\mathbf{F}_{k+1}\|^2.\end{aligned}$$

This proves the lemma. □

Note that by using applying Theorem 2.4.1 to ${F_k}^*\omega_k$ we get

$$\int_M \mathrm{Ric}({F_k}^*\omega_k) = 2g - 2 - \beta_{k+1}.$$

This, together with the above lemma, gives the following:

Theorem 2.5.3 (Plücker formula). *Assume that f is linearly nondegenerate. Then*

$$\int_M F_{k-1}{}^* \omega_{k-1} - \int_M 2F_k{}^* \omega_k + \int_M F_{k+1}{}^* \omega_{k+1} = 2g - 2 - \beta_{k+1}.$$

Now we extend the degree formula (2.12). Let f be a nonconstant holomorphic map of a compact Riemann surface M into $\mathbb{P}^n(\mathbb{C})$. The degree of f, denoted by $\deg(f)$, is the number of points of $f^{-1}(H)$, counting multiplicities, for a (and any) hyperplane H in $\mathbb{P}^n(\mathbb{C})$ such that $f(\mathbb{C}) \not\subset H$. We prove that

$$\deg(f) = \int_M f^* \omega_{\text{FS}}. \tag{2.17}$$

Indeed, let $H = \{a_0 z_0 + \cdots + a_n z_n = 0\}$ be such that $f(\mathbb{C}) \not\subset H$. Without loss of generality, we assume that $H = \{z_0 = 0\}$. Let $p_l, 1 \leq l \leq k$, be the (distinct) zeros of f_0, where $\mathbf{f} = (f_0, \ldots, f_n)$ is the reduced representation of f. Choose holomorphic local coordinates z_l with $z_l(p_l) = 0$. We consider a sufficiently small positive ϵ such that $\overline{U}_l(\epsilon) := \{z_l \mid |z_l| \leq \epsilon\}$ are mutually disjoint. Noticing that $f^* \omega_{\text{FS}} = dd^c \log \|\mathbf{f}\|^2$ and that $dd^c \log |f_0|^2 = 0$ outside $\{p_0, \ldots, p_k\}$, we obtain

$$\begin{aligned}
\int_M f^* \omega_{\text{FS}} &= \lim_{\epsilon \to 0} \int_{M - \bigcup_{l=1}^k U_l(\epsilon)} f^* \omega_{\text{FS}} \\
&= \lim_{\epsilon \to 0} \int_{M - \bigcup_{l=1}^k U_l(\epsilon)} dd^c \log\Big(1 + \sum_{i=1}^n |f_i/f_0|^2\Big) \\
&= -\lim_{\epsilon \to 0} \sum_{l=1}^k \int_{\partial U_l(\epsilon)} d^c \log\Big(1 + \sum_{i=1}^n |f_i/f_0|^2\Big) \\
&= -\lim_{\epsilon \to 0} \sum_{l=1}^k \int_{\partial U_l(\epsilon)} d^c \log(|f_0|^2 + \cdots + |f_n|^2) \\
&\quad + \lim_{\epsilon \to 0} \sum_{l=1}^k \int_{\partial U_l(\epsilon)} d^c \log |f_0|^2 = \lim_{\epsilon \to 0} \sum_{l=1}^k \int_{\partial U_l(\epsilon)} d^c \log |f_0|^2.
\end{aligned}$$

On the other hand,

$$\sum_{l=1}^k \int_{\partial U_l(\epsilon)} d^c \log |f_0|^2 = \sum_{l=1}^k \frac{1}{2\pi i} \int_{|z_l| = \epsilon} \frac{f_0'(z)}{f_0(z)} dz = \sum_{l=1}^k \nu_{f_0}(p_l) = \deg(f).$$

This proves (2.17).

Denote by $d_k = \deg(F_k)$. Then the Plücker formula becomes

$$d_{k-1} - 2d_k + d_{k+1} = 2g - 2 - \beta_{k+1}.$$

Using the fact that (assuming that $d_{-1} = 0$ and noticing that $d_n = 0$)

$$\sum_{k=0}^{n-1}(n-k)(d_{k-1} - 2d_k + d_{k+1}) = -(n+1)d_0,$$

we derived the following:

Theorem 2.5.4 (Brill–Segre formula). *Let $f : M \to \mathbb{P}^n(\mathbb{C})$ be a linearly nondegenerate holomorphic map, where M is a compact Riemann surface. Then*

$$\sum_{k=0}^{n-1}(n-k)\beta_{k+1} = n(n+1)(g-1) + (n+1)\deg(f).$$

A hyperplane H in $\mathbb{P}^n(\mathbb{C})$ is a subset

$$H = \{[z_0 : \cdots : z_n] \mid a_0 z_0 + \cdots + a_n z_n = 0\} \subset \mathbb{P}^n(\mathbb{C}).$$

The hyperplane is associated with the linear form $L(z_0, \ldots, z_n) = a_0 z_0 + \cdots + a_n z_n$. The hyperplanes $H_1, \ldots, H_q$ in $\mathbb{P}^n(\mathbb{C})$ are said to be *in general position* if for any $I \subset \{1, \ldots, q\}$ with $\#I \leq n+1$, $L_i, i \in I$, are linearly independent over $\mathbb{C}$, where $L_j, 1 \leq j \leq q$, are the linear forms defining H_j.

Theorem 2.5.5 (SMT for linearly nondegenerate algebraic curves, simple version). *Let M be a compact Riemann surface of genus g. Let $f : M \to \mathbb{P}^n(\mathbb{C})$ be a linearly nondegenerate holomorphic map. Let $H_1, \ldots, H_q$ be the hyperplanes in $\mathbb{P}^n(\mathbb{C})$ located in general position. Let $E = \bigcup_{j=1}^{q} f^{-1}(H_j)$. Then*

$$(q - (n+1))\deg(f) \leq \frac{1}{2}n(n+1)\{2(g-1) + |E|\}.$$

Proof. Let $P \in E$. Take a local reduced representation $\mathbf{f} = (f_0, \ldots, f_n)$ of f with the normal form around P, that is,

$$f_0 = t^{\delta_0(P)} + \cdots, \ldots, f_n = t^{\delta_n(P)} + \cdots$$

with $0 = \delta_0(P) < \delta_1(P) < \cdots < \delta_n(P)$, where t is a local parameter around P. Let $L_j, 1 \leq j \leq q$, be the linear forms defining H_j. Denote by $l_j := L_j(\mathbf{f})$. For each point $P \in E$, denote by $v_{l_j}(P)$ the vanishing order of l_j at P. Since $L_j, 1 \leq j \leq q$, are in general position, we can choose $\{L_{P,j_1}, \ldots, L_{P,j_n}\} \subset \{L_1, \ldots, L_q\}$ such that $v_{l_{P,j}}(P) = 0$ for $j \neq j_1, \ldots, j_n$ and

$$v_{l_{P,j_1}}(P) \leq v_{l_{P,j_2}}(P) \leq \cdots \leq v_{l_{P,j_n}}(P).$$

We claim that for $k = 1, \dots, n$,

$$v_{l_{P,j_k}}(P) \le \delta_k(P).$$

Indeed, assume that the claim is false, that is, $v_{l_{P,j_k}}(P) > \delta_k(P)$ for some $1 \le k \le n$. Write $L_{P,j_\lambda} := a_{0,\lambda}w_0 + \cdots + a_{n,\lambda}w_n$ for $\lambda = 1, \dots, n$. Since

$$l_{P,j_\lambda} = L_{P,j_\lambda}(\mathbf{f}) = a_{0,\lambda}(t^{\delta_0(P)} + \cdots) + \cdots + a_{n,\lambda}(t^{\delta_n(P)} + \cdots),$$

by expanding both sides as power series in t and comparing the coefficients of them, as well as noticing that $v_{l_{P,j_1}}(P) \le v_{l_{P,j_2}}(P) \le \cdots \le v_{l_{P,j_n}}(P)$, we easily see that

$$a_{0,\lambda} = a_{1,\lambda} = \cdots = a_{k,\lambda} = 0$$

for $\lambda = k, \dots, n$. This implies that the linear forms $L_{P,j_k}, \dots, L_{P,j_n}$ are linearly dependent, contradicting with the "in general position" assumption. This proves the claim.

From (2.14) we have

$$\nu_k(P) = \delta_k(P) - \delta_{k-1}(P) - 1, \; k = 1, \dots, n,$$

and thus

$$\begin{aligned}\sum_{k=0}^{n-1}(n-k)\nu_{k+1}(P) &= \sum_{k=1}^{n}\delta_k(P) - \frac{1}{2}n(n+1)\\ &\ge \sum_{t=1}^{n} v_{l_{P,j_t}}(P) - \frac{1}{2}n(n+1) = \sum_{j=1}^{q} v_{l_j}(P) - \frac{1}{2}n(n+1),\end{aligned} \tag{2.18}$$

where the last identity holds because $\nu_{l_{P,j}}(P) = 0$ for $j \ne j_1, \dots, j_n$. Applying the Brill–Segre formula gives

$$\begin{aligned}\sum_{j=1}^{q}\sum_{P\in E} v_{l_j}(P) &\le \sum_{k=0}^{n-1}(n-k)\beta_{k+1} + \frac{1}{2}n(n+1)|E|\\ &= n(n+1)(g-1) + (n+1)\deg(f) + \frac{1}{2}n(n+1)|E|.\end{aligned}$$

Since $\sum_{P\in E} v_{l_j}(P) = \deg(f)$ for each j, we get

$$(q-(n+1))\deg(f) \le \frac{1}{2}n(n+1)\{2(g-1) + |E|\},$$

which proves the theorem. □

The theorem however is not sufficient to serve our purpose. We need to modify the above proof to derive a general second main theorem. In the theorem below, we do not take $E = \bigcup_{j=1}^{q} f^{-1}(H_j)$; we rather let E be an arbitrary finite subset of M.

Theorem 2.5.6 (The SMT for algebraic curves with truncation). *Let M be a compact Riemann surface of genus g. Let $f : M \to \mathbb{P}^n(\mathbb{C})$ be a holomorphic map. Assume that $f(M)$ is contained in some k-dimensional projective subspace of $\mathbb{P}^n(\mathbb{C})$ but not in any subspace of dimension lower than k, where $1 \leq k \leq n$. Let $H_1, \ldots, H_q$ be the hyperplanes in $\mathbb{P}^n(\mathbb{C})$ located in general position, and let $L_1, \ldots, L_q$ be the corresponding linear forms. Assume that $L_j(\mathbf{f}) \not\equiv 0$ for $1 \leq j \leq q$. Let S be a finite subset of M. Then*

$$(q - 2n + k - 1)\deg(f) \leq \sum_{j=1}^{q} \sum_{P \notin S} \min\{k, v_{L_j(f)}(P)\} + \frac{1}{2}k(2n - k + 1)\{2(g - 1) + |S|\},$$

where $v_{L_j(f)}(P)$ is the vanishing order of $L_j(f)$ at the point P.

Proof. We first consider the case that f is linearly nondegenerate. We only need to modify the proof given above. As above, for each $P \in M$, we can choose $\{L_{P,j_1}, \ldots, L_{P,j_n}\} \subset \{L_1, \ldots, L_q\}$ such that $v_{l_{P,j}}(P) = 0$ for $j \neq j_1, \ldots, j_n$ and

$$v_{l_{P,j_1}}(P) \leq v_{l_{P,j_2}}(P) \leq \cdots \leq v_{l_{P,j_n}}(P).$$

For $P \in S$, similarly to (2.18), we have

$$\sum_{k=0}^{n-1}(n - k)\nu_{k+1}(P) \geq \sum_{j=1}^{q} v_{l_j}(P) - \frac{1}{2}n(n + 1). \tag{2.19}$$

For $P \notin S$, if we let $I_P = \{l_{P,j_\lambda}; v_{l_{P,j_\lambda}}(P) > 0, 1 \leq \lambda \leq n\}$ and use (2.18), then

$$\begin{aligned}
\sum_{k=0}^{n-1}(n - k)\nu_{k+1}(P) &\geq \sum_{t=1}^{n}(v_{l_{P,j_t}}(P) - t) \\
&\geq \sum_{l_{j_\lambda} \in I_P} \max\{0, v_{l_{P,j_\lambda}}(P) - n\} \\
&= \sum_{l_{j_\lambda} \in I_P} (v_{l_{P,j_\lambda}}(P) - \min\{n, v_{l_{P,j_\lambda}}(P)\}) \\
&= \sum_{t=1}^{n}(v_{l_{P,j_t}}(P) - \min\{n, v_{l_{P,j_t}}(P)\}) \\
&= \sum_{j=1}^{q}(v_{l_j}(P) - \min\{n, v_{l_j}(P)\}).
\end{aligned}$$

Combining this inequality with (2.19), we get

$$\sum_{k=0}^{n-1}(n - k)\beta_{k+1} = \sum_{k=0}^{n-1}\sum_{P \in S}(n - k)\nu_{k+1}(P) + \sum_{k=0}^{n-1}\sum_{P \notin S}(n - k)\nu_{k+1}(P)$$

$$\geq \sum_{j=1}^{q} \sum_{P \in M} v_{l_j}(P) - \sum_{j=1}^{q} \sum_{P \notin S} \min\{n, v_{l_j}(P)\} - \frac{n(n+1)}{2}|S|$$

$$= q \deg(f) - \sum_{j=1}^{q} \sum_{P \notin S} \min\{n, v_{l_j}(P)\} - \frac{n(n+1)}{2}|S|,$$

where we used the fact that $\sum_{p \in M} v_{l_j}(P) = \deg(f)$ for each j. Applying this to the Brill–Segre formula proves the theorem.

The degenerate case can be done by using the Nochka weights (see Section 4.3). We omit the details. □

3 Minimal surfaces in $\mathbb{R}^3$

In this chapter, we present the theory of value distirbutions of the Gauss maps of minimal surfaces in $\mathbb{R}^3$. For the references, see [63, 34, 3, 16].

3.1 The minimal surfaces

A surface in $\mathbb{R}^3$ is called a *minimal* surface if its mean curvature is identically zero. Recall that the mean curvature $H = \frac{1}{2}(\kappa_1 + \kappa_2)$, where κ_1 and κ_2 are the principal curvatures. So, geometrically, a minimal surface has the following character: at each point, the two principal curvatures (the largest and smallest normal curvatures) are canceled out. Also note that the Gauss curvature K of a minimal surface is always nonpositive, since $K = \kappa_1\kappa_2 = -\kappa_1^2 \leq 0$ when $H = \frac{1}{2}(\kappa_1 + \kappa_2) = 0$. Furthermore, if $K \equiv 0$, then $\kappa_1 = \kappa_2 = 0$, hence all normal curvatures are zero, and therefore the surface lies in a plane.

Theorem 3.1.1. *For a minimal surface M immersed in $\mathbb{R}^3$, its Gauss curvature $K \equiv 0$ if and only if it lies in a plane.*

The followings are some standard examples of minimal surfaces.

– The *catenoid* is given by

$$\mathbf{x}(u, v) = (a \cosh v \cos u, a \cosh v \sin u, av), \quad 0 < u < 2\pi, -\infty < v < +\infty.$$

This is a surface of revolution generated by rotating the catenary $y = a\cosh(\frac{z}{a})$ about the z-axis. It is easily checked that its mean curvature $H \equiv 0$. Thus the catenoid is a minimal surface. It can be characterized as the only surface of revolution that is minimal, that is, if a surface of revolution M is a minimal surface, then M is contained in either a plane or a catenoid.

– A *helicoid* is a ruled surface swept out by a straight line that rotates at constant speed about an axis perpendicular to the line while simultaneously moving at constant speed along the axis. We can take the axis to be the z-axis. Let ω be the angular velocity of the rotating line, and let a be its speed along the z-axis. If the line starts along the x-axis, at time v the center of the line is at $(0, 0, av)$, and it has rotated by the angle ωv. Hence the point of the line initially at $(u, 0, 0)$ is now at the point with position vector

$$\mathbf{x}(u, v) = (u \cos \omega v, u \sin \omega v, av).$$

This is a parameterization of the *helicoid*. It is easily checked that its mean curvature $H \equiv 0$. Thus the helicoid is a minimal surface. It can be characterized as the only minimal surface, other than the plane, which is also a ruled surface; that is, any ruled minimal surface in $\mathbb{R}^3$ is part of a plane or a helicoid.

https://doi.org/10.1515/9783110989557-003

- *Enneper's* surface is defined by

$$\mathbf{x}(u,v) = \left(u - \frac{1}{3}u^3 + uv^2, v - \frac{1}{3}v^3 + vu^2, u^2 - v^2\right).$$

  It is easily checked that its mean curvature $H \equiv 0$. Thus Enneper's surface is a minimal surface.
- *Scherk's* surface is defined by the Cartesian equation

$$z = \ln\left(\frac{\cos y}{\cos x}\right).$$

- The graph $z = f(x,y)$. In this case the parameterization is $\mathbf{x}(x,y) = (x,y,f(x,y))$. Hence the coefficients of the first fundamental form are $g_{11} = 1 + f_x^2, g_{12} = f_x f_y, g_{22} = 1 + f_y^2$, and the coefficients of the second fundamental form are

$$h_{11} = \frac{f_{xx}}{\sqrt{1 + f_x^2 + f_y^2}}, \quad h_{12} = \frac{f_{xy}}{\sqrt{1 + f_x^2 + f_y^2}}, \quad h_{22} = \frac{f_{yy}}{\sqrt{1 + f_x^2 + f_y^2}}.$$

  Therefore

$$H = \frac{f_{xx}(1 + f_y^2) - 2f_x f_y f_{xy} + (1 + f_x^2) f_{yy}}{2(1 + f_x^2 + f_y^2)^{3/2}}.$$

  The graph $z = f(x,y)$ is a minimal surface if and only if $H = 0$, that is,

$$(1 + f_y^2) f_{xx} - 2f_x f_y f_{xy} + (1 + f_x^2) f_{yy} = 0.$$

We will now try to understand why the word "minimal" is used for such surfaces. Consider the problem first posed in the mid-1800s by the Belgian physicist Plateau: *given a curve C, find a minimal surface M having C as its boundary*. As we will see, least-area surfaces are minimal. Thus another version of Plateau's problem is to find a least-area surface having C as its boundary.

What is a necessary condition that M have least-area among all surfaces with boundary C? The answer may be found in a simplified version of the calculus of variations as follows: Let M be the surface with parameterization $\mathbf{x} : U \to \mathbb{R}^3$. Let γ be a curve in U that bounds a subdomain $\triangle \subset U$. Consider the nearby surfaces that look like slightly deformed versions of M,

$$M_\lambda : \mathbf{g}(u,v) = \mathbf{x}(u,v) + \lambda h(u,v)\mathbf{n}(u,v), \quad (u,v) \in \triangle,$$

where λ is a real number, $\mathbf{n}(u,v)$ is the unit normal to the surface M, and h is a C^2 function in U that has the effect, when multiplied by a small λ and added to $\mathbf{x}$, of moving points of M a small bit and leaving the curve γ fixed, that is, h vanishes on the curve γ. The surface M_λ is called a *normal variation* of M since we are varying the surface M via the parameter λ along the direction of our unit normal $\mathbf{n}$. Denoting by A_λ the area of the surface M_λ, we will prove the following:

Theorem 3.1.2.

$$\left.\frac{dA}{d\lambda}\right|_{\lambda=0} = -2\iint\limits_{\triangle} hH(g_{11}g_{22} - g_{12}^2)^{1/2} h\,du\,dv,$$

where H is the mean curvature of M, and $\triangle$ is the interior of γ.

Proof. Differentiating $\mathbf{g}$ with respect to u and v, we have

$$\mathbf{g}_u = \mathbf{x}_u + \lambda(h\mathbf{n}_u + h_u\mathbf{n}), \quad \mathbf{g}_v = \mathbf{x}_v + \lambda(h\mathbf{n}_v + h_v\mathbf{n}).$$

If we let $g_{11}^\lambda, g_{12}^\lambda, g_{22}^\lambda$ be the first fundamental form of the surface M_λ, then from the above we get

$$g_{11}^\lambda = g_{11} - 2\lambda h h_{11} + O(\lambda^2), \quad F^\lambda = F - 2\lambda h h_{12} + O(\lambda^2),$$

and

$$g_{22}^\lambda = g_{22} - 2\lambda h h_{22} + O(\lambda^2),$$

where h_{11}, h_{12}, h_{22} are the coefficients of the second fundamental form, and $O(\lambda^2)$ means λ^2 times some continuous function. So we have

$$g_{11}^\lambda g_{22}^\lambda - g_{12}^{\lambda\,2} = a_0 + a_1\lambda + O(\lambda^2),$$

where $a_0 = g_{11}g_{22} - g_{12}^2$, $a_1 = -2h(g_{11}h_{22} + g_{22}h_{11} - 2g_{12}h_{12})$. This implies that for $(u, v) \in \triangle$,

$$\left|(E^\lambda G^\lambda - F^{\lambda 2})^{1/2} - \left(\sqrt{a_0} + \frac{a_1}{2\sqrt{a_0}}\lambda\right)\right| < C_1\lambda^2,$$

where $C_1 > 0$ is a constant. By the formulas

$$A_\lambda = \iint\limits_{\triangle} (g_{11}^\lambda g_{22}^\lambda - g_{12}^{\lambda\,2})^{1/2} du\,dv$$

and

$$A_0 = \iint\limits_{\triangle} (g_{11}g_{22} - g_{12}^2)^{1/2} du\,dv = \iint\limits_{\triangle} \sqrt{a_0}\,du\,dv$$

we get

$$\left|A_\lambda - A_0 - \lambda \iint\limits_{\triangle} \frac{a_1}{2\sqrt{a_0}} dudv\right| < C_1\lambda^2.$$

This implies that

$$\left|\frac{A_\lambda - A_0}{\lambda} - \iint\limits_{\triangle} \frac{a_1}{2\sqrt{a_0}} dudv\right| < C_1\lambda.$$

Letting λ go to zero and using $H = \frac{1}{2}\frac{g_{11}h_{22}-2g_{12}h_{12}+g_{22}h_{11}}{g_{11}g_{22}-g_{12}^2}$, we get

$$A'(0) = -2\iint\limits_{\triangle} hH(g_{11}g_{22} - g_{12}^2)^{1/2} hdudv.$$

This proves the theorem. □

Theorem 3.1.3. *Let $\mathbf{x} : U \to \mathbb{R}^3$ be a regular parameterized surface, and let $D \subset U$ be a bounded domain in U. Then $\mathbf{x}$ is minimal on D if and only if $A'(0) = 0$ for all such D and all normal variations of $\mathbf{x}(\overline{D})$.*

Proof. If $\mathbf{x}$ is minimal, then H is identically zero, and so $A'(0) = 0$ for any h. Conversely, assume that $A'(0) = 0$ for any h but that there exists $q \in D$ for which its mean curvature $H(q) \neq 0$. We choose $h : \overline{D} \to \mathbb{R}$ such that $h(q) = H(q)$ and h is identically zero outside a small neighborhood of q. Then $A'(0) < 0$ for the variation determined by this h. This contradiction shows that $H(q) = 0$. Since q is arbitrary, $\mathbf{x}$ is minimal. □

Note that we have said nothing about the second derivative of A at 0, so that a minimal surface, although a critical point of A, may not actually be a minimum.

3.2 The Weierstrass–Enneper representation

Let M be a surface in $\mathbb{R}^3$ with parameterization $\mathbf{x} : U \to \mathbb{R}^3$. The parameterization $\mathbf{x}$ is said to be *isothermal* if

$$\|\mathbf{x}_u\| = \|\mathbf{x}_v\| \quad \text{and} \quad \mathbf{x}_u \cdot \mathbf{x}_v = 0.$$

Thus $\mathbf{x}$ is a conformal mapping (i. e., it preserves angles), and its induced metric from the standard Euclidean metric in $\mathbb{R}^3$ is $ds^2 = \lambda^2(du^2 + dv^2)$, where $\lambda = \|\mathbf{x}_u\| = \|\mathbf{x}_v\|$.

Lemma 3.2.1 (Existence of isothermal parameters on a minimal surface). *Let M be a minimal surface in $\mathbb{R}^3$. Then there is a (local) isothermal parameterization of M.*

Proof. Since it is a local problem, we can assume that M is the graph of some function, so we have a local parameterization of $\boldsymbol{\sigma}(x,y) = (x, y, f(x,y)) : U \subset \mathbb{R}^2 \to \mathbb{R}^3$ for M. The coefficients of the first fundamental form are

$$g_{11} = 1 + f_x^2, \quad g_{12} = f_x f_y, \quad g_{22} = 1 + f_y^2.$$

The graph $z = f(x, y)$ is a minimal surface if and only if $H = 0$, which means

$$(1 + f_y^2) f_{xx} - 2 f_x f_y f_{xy} + (1 + f_x^2) f_{yy} = 0. \tag{3.1}$$

Let $A = \sqrt{g_{11} g_{22} - g_{12}^2} = \sqrt{1 + f_x^2 + f_y^2}$. It can easily be checked through a direct computation by using (3.1) that

$$\frac{\partial}{\partial x}\left(\frac{g_{12}}{A}\right) = \frac{\partial}{\partial y}\left(\frac{g_{11}}{A}\right) \tag{3.2}$$

$$\frac{\partial}{\partial x}\left(\frac{g_{22}}{A}\right) = \frac{\partial}{\partial y}\left(\frac{g_{12}}{A}\right). \tag{3.3}$$

For example, to verify (3.2), we proceed as follows:

$$\begin{aligned}
&\frac{\partial}{\partial x}\left(\frac{f_x f_y}{\sqrt{1 + f_x^2 + f_y^2}}\right) - \frac{\partial}{\partial y}\left(\frac{1 + f_x^2}{\sqrt{1 + f_x^2 + f_y^2}}\right) \\
&= \frac{(1 + f_x^2 + f_y^2)(f_{xx} f_y + f_x f_{xy}) - f_x f_y (f_x f_{xx} + f_y f_{yx})}{(1 + f_x^2 + f_y^2)^{3/2}} \\
&\quad - \frac{2(1 + f_x^2 + f_y^2) f_x f_{xy} - (1 + f_x^2)(f_x f_{xy} + f_y f_{yy})}{(1 + f_x^2 + f_y^2)^{3/2}} \\
&= \frac{f_y [(1 + f_y^2) f_{xx} - 2 f_x f_y f_{xy} + (1 + f_x^2) f_{yy}]}{(1 + f_x^2 + f_y^2)^{3/2}} = 0.
\end{aligned}$$

Thus there are smooth functions ϕ and ψ on $U \subset \mathbb{R}^2$ such that

$$\phi_x = \frac{g_{11}}{A}, \quad \phi_y = \frac{g_{12}}{A}, \quad \psi_x = \frac{g_{12}}{A}, \quad \psi_y = \frac{g_{22}}{A}.$$

Set $L(x, y) = (x + \phi(x, y), y + \psi(x, y)) = (u, v) : U \subset \mathbb{R}^2 \to \mathbb{R}^2$ (here L is called the Levy map). Then the Jacobian matrix of L is

$$J = \begin{pmatrix} 1 + \frac{g_{11}}{A} & \frac{g_{12}}{A} \\ \frac{g_{12}}{A} & 1 + \frac{g_{22}}{A} \end{pmatrix}, \tag{3.4}$$

and hence

$$\det(J) = 2 + \frac{g_{11} + g_{22}}{A} > 0.$$

Therefore by the inverse function theorem the inverse map L^{-1} exists. Let $\tilde{\boldsymbol{\sigma}}(u, v) := (\boldsymbol{\sigma} \circ L^{-1})(u, v)$. By the chain rule,

$$\begin{pmatrix} \tilde{\boldsymbol{\sigma}}_u \\ \tilde{\boldsymbol{\sigma}}_v \end{pmatrix} = \begin{pmatrix} x_u & y_u \\ x_v & y_v \end{pmatrix} \begin{pmatrix} \boldsymbol{\sigma}_x \\ \boldsymbol{\sigma}_y \end{pmatrix}.$$

Hence

$$\begin{pmatrix} \tilde{\boldsymbol{\sigma}}_u \\ \tilde{\boldsymbol{\sigma}}_v \end{pmatrix}^t \cdot \begin{pmatrix} \tilde{\boldsymbol{\sigma}}_u \\ \tilde{\boldsymbol{\sigma}}_v \end{pmatrix} = J^{-1} \begin{pmatrix} g_{11} & g_{12} \\ g_{12} & g_{22} \end{pmatrix} (J^{-1})^t = \frac{A}{\det J} \begin{pmatrix} 1 & 0 \\ 0 & 1 \end{pmatrix}.$$

Hence we get

$$\tilde{\boldsymbol{\sigma}}_u \cdot \tilde{\boldsymbol{\sigma}}_u = \tilde{\boldsymbol{\sigma}}_v \cdot \tilde{\boldsymbol{\sigma}}_v = \frac{A}{\det J}, \quad \tilde{\boldsymbol{\sigma}}_u \cdot \tilde{\boldsymbol{\sigma}}_v = 0. \tag{3.5}$$

This shows that the new parameterization $\tilde{\boldsymbol{\sigma}}$ is isothermal. □

Let M be a regular minimal surface in $\mathbb{R}^3$ of class C^2. By the theorem above each point $p \in M$ has a neighborhood in which an isothermal parameter $\mathbf{x}(u, v)$ is defined.

Proposition 3.2.2. *For an oriented surface M with metric ds^2, if we take two systems of positively oriented isothermal local parameters (or coordinates (u, v) and (x, y)), then $w := u + \sqrt{-1}v$ is a holomorphic function in $z := x + \sqrt{-1}y$ on the common domain of definition.*

Proof. By assumption there exists a positive differentiable function λ such that

$$ds^2 = \lambda^2(dx^2 + dy^2).$$

Therefore we have

$$A := u_x^2 + v_x^2 = u_y^2 + v_y^2, \quad u_x u_y + v_x v_y = 0.$$

This means that the Jacobi matrix

$$J := \begin{pmatrix} u_x & v_x \\ u_y & v_y \end{pmatrix}$$

satisfies the identity $J^t J = A^2 I_2$, where I_2 is the 2×2 identity matrix. We then have

$$J^{-1} = \frac{1}{\det J} \begin{pmatrix} v_y & -v_x \\ -u_y & u_x \end{pmatrix} = \frac{1}{A} \begin{pmatrix} v_y & -v_x \\ -u_y & u_x \end{pmatrix}.$$

On the other hand, since $(\det J)^2 = \det(J^t J) = \det(A^2 I_2) = A^2$ and $\det J > 0$, we have $A = \det J$. This implies that $u_x = u_y$ and $u_y = -v_x$. Therefore the function $w = u + iv$ is holomorphic in z. □

From the above proposition assume that M is minimal and orientable. Then M is a Riemann surface with a family of local isothermal parameters that covers M, since such changes of coordinates are holomorphic with $z = u + \sqrt{-1}v$. Let $\mathbf{x} : M \to \mathbb{R}^3$ be a minimal immersion, and let $\{U, \phi = (u, v)\}$ be a local isothermal coordinate. Denote $\mathbf{x}(u, v) = (\mathbf{x} \circ \phi^{-1})(u, v)$, which is a local isothermal parameterization. The evaluation of many quantities in the study of surfaces are simplified when considered in reference to the Riemann surfaces. For example, the induced metric is $ds^2 = \lambda^2(du^2 + dv^2) = \lambda^2 dzd\bar{z}$, where $z = u + \sqrt{-1}v$. We denote by $\triangle$, or more precisely $\triangle_z$, the Laplacian operator $\frac{\partial^2}{\partial u^2} + \frac{\partial^2}{\partial v^2}$ in terms of the holomorphic coordinate $z = u + iv$. If we take another local coordinate ζ, then we have $\triangle_\zeta = |dz/d\zeta|^2 \triangle_z$. Thus the operator $\triangle_{ds^2} := (1/\lambda^2)\triangle_z$ is globally defined (independent of the choice of the holomorphic coordinate z) and is called the *Laplacian operator with respect to the metric* ds^2. Denote

$$\triangle_{ds^2} := \frac{1}{\lambda^2}\left(\frac{\partial}{\partial u^2} + \frac{\partial}{\partial v^2}\right) = \frac{4}{\lambda^2}\frac{\partial^2}{\partial z \partial \bar{z}}.$$

The Gauss curvature by (2.7) is

$$K = -\triangle_{ds^2} \log \lambda. \tag{3.6}$$

For an immersion $\mathbf{x} = (x_1, x_2, x_3) : M \to \mathbb{R}^3$, we define

$$\triangle_{ds^2}\mathbf{x} = (\triangle_{ds^2}x_1, \triangle_{ds^2}x_2, \triangle_{ds^2}x_3).$$

We have

$$\triangle_{ds^2}\,\mathbf{x} = 2H\mathbf{n}. \tag{3.7}$$

To prove this identity, notice that $\mathbf{x}(u, v)$ is isothermal, so $\mathbf{x}_u \cdot \mathbf{x}_u = \mathbf{x}_v \cdot \mathbf{x}_v$ and $\mathbf{x}_u \cdot \mathbf{x}_v = 0$. By differentiating these equations with respect to u and v separately we get

$$\mathbf{x}_{uu} \cdot \mathbf{x}_u = \mathbf{x}_{uv} \cdot \mathbf{x}_v \quad \text{and} \quad \mathbf{x}_{vu} \cdot \mathbf{x}_v = -\mathbf{x}_{vv} \cdot \mathbf{x}_u.$$

Hence

$$(\mathbf{x}_{uu} + \mathbf{x}_{vv}) \cdot \mathbf{x}_u = \mathbf{x}_{uv} \cdot \mathbf{x}_v - \mathbf{x}_{vu} \cdot \mathbf{x}_v = 0.$$

Similarly, $(\mathbf{x}_{uu} + \mathbf{x}_{vv}) \cdot \mathbf{x}_v = 0$. From this we conclude that $\triangle_{ds^2}\mathbf{x}$ is proportional to $\mathbf{n}$. Now write $\mathbf{n}_u = a_{11}\mathbf{x}_u + a_{21}\mathbf{x}_v$ and $\mathbf{n}_v = a_{12}\mathbf{x}_u + a_{22}\mathbf{x}_v$. Since $H = \frac{1}{2}\mathrm{trace}(S_P) = -\frac{1}{2}\mathrm{trace}(d\mathbf{n}|_P)$, we have

$$H = -\frac{1}{2}(a_{11} + a_{22}).$$

Hence

$$\begin{aligned}\lambda^2(\triangle_{ds^2}\mathbf{x}) \cdot \mathbf{n} &= (\mathbf{x}_{uu} + \mathbf{x}_{vv}) \cdot \mathbf{n} \\ &= -\mathbf{x}_u \cdot \mathbf{n}_u - \mathbf{x}_v \cdot \mathbf{n}_v = -a_{11}\|\mathbf{x}_u\| - a_{22}\|\mathbf{x}_v\|\end{aligned}$$

$$= -(a_{11} + a_{22})\lambda^2 = 2H\lambda^2.$$

This proves (3.7).

Therefore we get the following proposition.

Proposition 3.2.3. *A surface M immersed in $\mathbb{R}^3$ is minimal if only if its isothermal parameterization $\mathbf{x}$ is harmonic.*

Corollary 3.2.4. *A minimal surface immersed in $\mathbb{R}^3$ cannot be compact.*

Proof. If M were compact, then each $x_k, 1 \le k \le 3$, would attain its maximum on M, where $\mathbf{x} = (x_1, x_2, x_3) : M \to \mathbb{R}^3$ is an immersion of M. Since by Proposition 3.2.3 each $x_k, 1 \le k \le 3$, is harmonic, it would be constant, contradicting the assumption that M is a surface. □

Let M be a minimal surface immersed in $\mathbb{R}^3$ with an immersion $\mathbf{x} : M \to \mathbb{R}^3$. We know that M is always noncompact. Take a local isothermal coordinate $\{U, \phi = (u, v)\}$ of M. Write $\mathbf{x}(u, v) = (\mathbf{x} \circ \phi^{-1})(u, v) = (x_1(u, v), x_2(u, v), x_3(u, v))$. Let $\phi = (\phi_1, \phi_2, \phi_3)$, where, for $1 \le k \le 3$,

$$\phi_k = \frac{\partial x_k}{\partial z} = \frac{1}{2}\left(\frac{\partial x_k}{\partial u} - \sqrt{-1}\frac{\partial x_k}{\partial v}\right).$$

The condition of M being minimal implies that $\mathbf{x}$ is harmonic, that is, $\frac{\partial^2 \mathbf{x}}{\partial z \partial \bar{z}} = \mathbf{0}$. Thus $\frac{\partial \phi_k}{\partial \bar{z}} = 0$, so ϕ_k is holomorphic for $k = 1, 2, 3$. Also,

$$\begin{aligned} 4\sum_{k=1}^{3} \phi_k^2 &= \sum_{k=1}^{3}\left(\frac{\partial x_k}{\partial u}\right)^2 - \sum_{k=1}^{3}\left(\frac{\partial x_k}{\partial v}\right)^2 - 2\sqrt{-1}\sum_{k=1}^{3} \frac{\partial x_k}{\partial u}\frac{\partial x_k}{\partial v} \\ &= \|\mathbf{x}_u\|^2 - \|\mathbf{x}_v\|^2 - 2\sqrt{-1}\mathbf{x}_u \cdot \mathbf{x}_v = 0, \end{aligned}$$

and

$$4\sum_{k=1}^{3} |\phi_k|^2 = \sum_{k=1}^{3}\left(\frac{\partial x_k}{\partial u}\right)^2 + \sum_{k=1}^{3}\left(\frac{\partial x_k}{\partial v}\right)^2 = \|\mathbf{x}_u\|^2 + \|\mathbf{x}_v\|^2 \neq 0$$

if M is regular. The differential $\alpha_k := \phi_k dz, 1 \le k \le 3$, is independent of the local coordinate z and hence is a globally defined differential form on M. Thus we obtain the following:

Proposition 3.2.5. *Let M be a Riemann surface. Then $\alpha = \phi dz$ is a vector-valued holomorphic form on M if and only if $\mathbf{x}$ is a minimal surface. Moreover,*

$$\mathbf{x} = 2\,\mathrm{Re}\left(\int \phi\right),$$

where the integral is taken along any path from a fixed point to z on M.

When the real part of the integral of α along any closed path is zero, we say that α has no real periods. The nonexistence of real periods for α is easily seen to be equivalent to the independence of (Re $\int^z \alpha$) on the path M.

Theorem 3.2.6 (Weierstrass–Enneper representation). *Let α_1, α_2, and α_3 be holomorphic differentials on M such that $\alpha_i = \phi_i dz$. Suppose that*

1. $\phi_1^2 + \phi_2^2 + \phi_3^2 = 0$;
2. $\sum_{k=1}^3 |\phi_k(z)|^2 > 0$;
3. *Each α_k has no real periods on M.*

Then the mapping $\mathbf{x} : M \to \mathbb{R}^3$ defined by $x_k(z) = \mathrm{Re}(\int^z \alpha_k)$ is a minimal immersion.

Condition (3) of the theorem is necessary to guarantee that $\mathrm{Re}(\int_{z_0}^z \alpha_k)$ depends only on the final point z. Therefore each x_k is well defined independently of the path from z_0 to z. It is obvious that $\phi = \frac{\partial \mathbf{x}}{\partial z}$ is holomorphic, and so $\mathbf{x}$ is harmonic. Hence $\mathbf{x}$ is a minimal surface. Condition (2) guarantees that $\mathbf{x}$ is regular, so it is an immersion.

It is possible to give a simple description of all solutions of the equation $\sum_{k=1}^3 \phi_k = 0$ on M. To do this, we assume that $\phi_1 \neq i\phi_2$. (If $\phi_1 = i\phi_2$, then $\phi_3 = 0$, so that the minimal surface is a plane.) Let us define the holomorphic form ω and the meromorphic function g as follows:

$$\omega = \alpha_1 - i\alpha_2 \quad \text{and} \quad g = \frac{\alpha_3}{\alpha_1 - i\alpha_2}.$$

Locally, write $\alpha_k = \phi_k dz$. Then $\omega = f dz$, where $f = \phi_1 - i\phi_2$ and $g := \frac{\phi_3}{\phi_1 - i\phi_2}$. Then f is a (locally defined) holomorphic function, and g is meromorphic on M, which is globally defined. We call the pair $\{\omega, g\}$ (or simply $\{f, g\}$) the *Weierstrass–Enneper representation* of the minimal surface M. The induced metric on the surface M is

$$ds^2 = 2\|\phi\|^2 dz d\bar{z} = |f|^2 (1 + |g|^2)^2 dz d\bar{z}, \tag{3.8}$$

and its Gauss curvature by (3.6) is

$$K = -\frac{2|g'|}{|f|^2 (1 + |g|^2)^4}. \tag{3.9}$$

Note that if z_0 is a point where g has a pole of order m. Then it is clear that ω (or f) must have a zero of order exactly $2m_0$ at z_0 for each α_k to be holomorphic. In terms of g and ω, we have

$$\alpha_1 = \frac{1}{2}(1 - g^2)\omega, \quad \alpha_2 = \frac{i}{2}(1 + g^2)\omega, \quad \alpha_3 = g\omega.$$

Therefore the minimal immersion $\mathbf{x} = (x_1, x_2, x_3)$ is given by

$$x_1(z) = \mathrm{Re} \int_{z_0}^{z} f(1 - g^2)d\zeta,$$

$$x_2(z) = \mathrm{Re} \int_{z_0}^{z} if(1 + g^2)d\zeta,$$

$$x_3(z) = 2\,\mathrm{Re} \int_{z_0}^{z} fgd\zeta,$$

where the integral is taken along an arbitrary path from the fixed point $z_0 \in D$ to the point z.

Conversely, let $D \subset \mathbb{C}$ be a simply connected domain, let f be a holomorphic function in D, and let g be a meromorphic function in D such that fg^2 is holomorphic in D. Let

$$\phi_1 = \frac{1}{2}f(1 - g^2), \quad \phi_2 = \frac{i}{2}f(1 + g^2), \quad \phi_3 = fg.$$

Then it is clear that ϕ_1, ϕ_2, ϕ_3 are holomorphic and

$$\begin{aligned}\phi_1^2 + \phi_2^2 + \phi_3^2 &= \frac{1}{4}f^2(1 - g^2)^2 - \frac{1}{4}f^2(1 + g^2)^2 + f^2g^2\\ &= \frac{1}{4}f^2(1 - 2g^2 + g^4) - \frac{1}{4}f^2(1 + 2g^2 + g^4) + f^2 + g^2\\ &= -f^2g^2 + f^2g^2 = 0.\end{aligned}$$

This gives an alternative version of the Weierstrass–Enneper representation.

Theorem 3.2.7 (Weierstrass–Enneper representation). *If D is a simply connected domain in the complex plane and if f is a holomorphic function in D and g is a meromorphic function in D such that fg^2 is holomorphic in D, then the minimal surface is defined by the parameterization $\mathbf{x}(z) = (x_1(z), x_2(z), x_3(z))$ where*

$$x_1(z) = \mathrm{Re} \int_{z_0}^{z} f(1 - g^2)d\zeta,$$

$$x_2(z) = \mathrm{Re} \int_{z_0}^{z} if(1 + g^2)d\zeta,$$

$$x_3(z) = 2\,\mathrm{Re} \int_{z_0}^{z} fgd\zeta,$$

where the integral is taken along an arbitrary path from the fixed point $z_0 \in D$ to the point z.

It can be shown that the catenoid and helicoid have representations of the forms $(f(z), g(z)) = (-\frac{1}{2}e^{-z}, -e^z)$ and $(f(z), g(z)) = (-\frac{i}{2}e^{-z}, -e^z)$, respectively. It can also be shown that the minimal surface called the Enneper surface has a representation of the form $(f(z), g(z)) = (1, z)$.

The function g has a special geometric meaning: g is the stereographic projection of $\mathbf{n}$ (the unit normal to the minimal surface) from the unit sphere to $\mathbb{C} \cup \{\infty\} = \mathbb{P}^1(\mathbb{C})$. Here the stereographic projection from the north pole is the mapping $St : S^2 \backslash \{(0,0,1)\} \to \mathbb{R}^2$ defined by

$$St(x_1, x_2, x_3) = \left(\frac{x_1}{(1-x_3)}, \frac{x_2}{(1-x_3)}, 0\right).$$

Note that the stereographic projection from the north pole is a bijective, smooth, and conformal mapping where it is defined. We now identify $\mathbb{R}^2$ with $\mathbb{C}$ and extend the map St to a bijective mapping $St : S^2 \to \mathbb{C} \cup \{\infty\}$ defined by

$$(x_1, x_2, x_3) \mapsto \frac{x_1 + \sqrt{-1}x_2}{1 - x_3} \quad \text{and} \quad (0,0,0) \mapsto \infty. \tag{3.10}$$

Theorem 3.2.8. *Let M be a minimal surface with isothermal parameterization $\mathbf{x}(u, v)$ and Weierstrass–Enneper representation (f, g). Then $g = St(\mathbf{n})$, that is, $g : M \to \mathbb{P}^1(\mathbb{C})$ can be regarded as the Gauss map of M given by $\mathbf{n} : M \to S^2$.*

Proof. Let $\mathbf{x} = (x_1, x_2, x_3)$ be a parameterization of M. To compute $\mathbf{n}$, we first compute $\mathbf{x}_u \times \mathbf{x}_v$. By a direct computation,

$$\mathbf{x}_u \times \mathbf{x}_v = ((x_2)_u(x_3)_v - (x_3)_u(x_2)_v, (x_3)_u(x_1)_v - (x_1)_u(x_3)_v, (x_1)_u(x_2)_v - (x_2)_u(x_1)_v).$$

Notice that

$$\begin{aligned}(x_2)_u(x_3)_v - (x_3)_u(x_2)_v &= \operatorname{Im}[((x_2)_u - i(x_2)_v)((x_3)_u + i(x_3)_v)] \\ &= \operatorname{Im}[2(\partial x_2/\partial z) \cdot 2(\partial x_3/\partial \bar{z})] \\ &= 4\operatorname{Im}(\phi_2\bar{\phi}_3).\end{aligned}$$

Similarly,

$$(x_3)_u(x_1)_v - (x_1)_u(x_3)_v = 4\operatorname{Im}(\phi_3\bar{\phi}_1)$$

and

$$(x_1)_u(x_2)_v - (x_2)_u(x_1)_v = 4\operatorname{Im}(\phi_1\bar{\phi}_2).$$

Hence

$$\mathbf{x}_u \times \mathbf{x}_v = 4\operatorname{Im}(\phi_2\bar{\phi}_3, \phi_3\bar{\phi}_1, \phi_1\bar{\phi}_2) = |f|^2(1 + |g|^2)(2\operatorname{Re}(g), 2\operatorname{Im}(g), (|g|^2 - 1)),$$

where we used $z - \bar{z} = 2\,\mathrm{Im}\,z$. Since $\mathbf{x}(u,v)$ is isothermal, $\|\mathbf{x}_u \times \mathbf{x}_v\| = \|\mathbf{x}_u\|\|\mathbf{x}_v\| = \|\mathbf{x}_u\|^2 = 2\|\phi\|^2 = |f|^2(1+|g|^2)^2$, it follows that

$$\mathbf{n} = \frac{\mathbf{x}_u \times \mathbf{x}_v}{\|\mathbf{x}_u \times \mathbf{x}_v\|} = \left(\frac{2\,\mathrm{Re}(g)}{1+|g|^2}, \frac{2\,\mathrm{Im}(g)}{1+|g|^2}, \frac{|g|^2-1}{1+|g|^2}\right).$$

Hence from (3.10) we have

$$St(\mathbf{n}) = \frac{\frac{2\,\mathrm{Re}(g)}{1+|g|^2} + \sqrt{-1}\frac{2\,\mathrm{Im}(g)}{1+|g|^2}}{1 - \frac{|g|^2-1}{1+|g|^2}} = g. \qquad \square$$

3.3 Gauss maps of complete minimal surfaces in $\mathbb{R}^3$

In this section, we prove the Bernstein's theorem and its extensions. For the references, see [4, 5, 6, 14, 18, 27, 30, 31, 32, 33, 42, 49, 55, 56, 57, 58, 59, 60, 61, 62, 87, 88, 89]. To do so, we need the following result regarding the flat metric on a Riemann surface.

Lemma 3.3.1. *Let $d\sigma^2$ be a conformal flat metric on an open surface M (i. e., its curvature is identically zero). Then for each point $p \in M$, there exists a local diffeomorphism Φ of a disk $D(R) = \{w \in \mathbb{C} \mid |w| < R\}(0 < R \le \infty)$ onto an open neighborhood of p in M with $\Phi(0) = p$ such that Φ is a local isometry (i. e., the pullback $\Phi^*(d\sigma^2)$ is equal to the standard Euclidean metric ds_E^2 on $D(R) \subset \mathbb{C}$), and there exists a point a_0 with $|a_0| = 1$ such that the Φ-image Γ_{a_0} of the line $L_{a_0} = \{w = a_0 t : 0 < t < R\}$ is divergent in M. In particular, if the metric $d\sigma^2$ is flat and complete, then M is biholomorphically isometric to the complex plane $\mathbb{C}$ with its standard flat metric $|dw|^2$.*

Proof. Write $d\sigma^2 = |f(z)|^2|dz|^2$. Then f is holomorphic and without zeros. Define

$$w = F(z) = \int_p^z f(\zeta)d\zeta,$$

where the integral $\int_p^z f(\zeta)d\zeta$ is over a simple path on M with initial point p and end-point z. Since f is holomorphic, the integral is independent of the choices of the paths, and hence F is well defined. Since $\frac{dw}{dz}|_p \ne 0$, F maps an open neighborhood U of p biholomorphically onto an open disc $D(R) := \{w \in \mathbb{C} \mid |w| < R\}(0 < R \le \infty)$. For the case $R = \infty$, we have $M = U$. In this case, every curve Γ_{a_0} is divergent. So we assume that $R < \infty$. Let R_0 be the least upper bound of $R > 0$ such that F biholomorphically maps some open neighborhood of p onto $D(R)$. By definition there is a sequence $\{R_n\}$ that converges to R_0 such that $F|_{U_n} : U_n \to D(R_n)$ is biholomorphic from some neighborhood U_n of p. Then F maps $U_0 = \bigcup_{n=1}^{\infty} U_n$ onto $D(R_0)$. Consider the inverse map $\Phi : D(R_0) \to U_0$. Assume that Γ_{a_0} is not divergent in M for some a_0 with $|a_0| = 1$. Then there is a sequence $\{t_n\}$ with $0 \le t_n < R_0$ such that $\lim_{n\to\infty} t_n = R_0$ and $\phi(t_n a_0)$ converges to a point $q \in M$.

Since $F(q) \neq 0$, F biholomorphically maps some open neighborhood of q onto an open neighborhood of $w_0 := R_0 a_0$. Therefore Φ is biholomorphically extended to a neighborhood of w_0. If there exists no curve Γ_{a_0} that diverges in M for all a_0 with $|a_0| = 1$, then Φ is biholomorphically extended to $D(R)$ for some $R > R_0$. This contradicts with the choice of R_0. Therefore some Γ_{a_0} diverges in M. Finally, we show that Φ is an isometry. Indeed,

$$\Phi^*(d\sigma^2) = |(f \circ \Phi)|^2 \left|\frac{dz}{dw}\right|^2 |dw|^2 = |dw|^2.$$

Since

$$\left|\frac{dz}{dw}\right| = \left|\frac{dw}{dz}\right|^{-1} = |(f \circ \Phi)|^{-1},$$

Φ is an isometry. □

Theorem 3.3.2 (Bernstein). *Let $M \subset \mathbb{R}^3$ be the graph $z = f(x,y)$. If M is minimal and f is defined on the whole plane $\mathbb{R}^2$, then $f(x,y) = ax + by + c$ (i. e., M is a plane).*

Proof. Let $\mathbf{x}(x,y) = (x,y,f(x,y)) : \mathbb{R}^2 \to \mathbb{R}^3$ be the standard parameterization of M. Take the new isothermal parameterization $\tilde{\mathbf{x}}(u,v) = (\mathbf{x} \circ L^{-1})(u,v)$ as described in Lemma 3.2.1, where L is the Levy map. The induced metric on M can be written as

$$ds^2 = \lambda^2(du^2 + dv^2),$$

where, according to (3.5),

$$\lambda^2 = \frac{1 + f_x^2 + f_y^2}{2\sqrt{1 + f_x^2 + f_y^2} + 2 + f_x^2 + f_y^2} = \frac{A^2}{2A + A^2 + 1} = \frac{A^2}{(A+1)^2} \tag{3.11}$$

with $A = \sqrt{1 + f_x^2 + f_y^2}$. Its Gauss curvature is, by formula (3.6),

$$K = \frac{-1}{\lambda^2}\left(\frac{\partial^2}{\partial u^2} + \frac{\partial^2}{\partial v^2}\right)\log\lambda = -\triangle_{ds^2}\log\lambda = \triangle_{ds^2}\log\left(1 + \frac{1}{A}\right). \tag{3.12}$$

We define the new metric

$$d\sigma^2 = \left(1 + \frac{1}{A}\right)^2 ds^2 = (du^2 + dv^2).$$

Then $d\sigma^2$ is a flat metric. Also, since $ds \leq d\sigma \leq 2ds$, we know that the metric $d\sigma^2$ on M is complete since ds^2 is complete. By Lemma 3.3.1, M with the metric $d\sigma^2$ is isometric to the complex plane $\mathbb{C}$ with its standard flat metric $d|w|$. On the other hand, since $K \leq 0$, by (3.12) the function $\psi := -\log(1 + \frac{1}{A})$, considered as a function on the $\mathbb{C}$-plane, is subharmonic. Clearly, $\psi \le 0$. By the well-known "Hadamard three-circle theorem" for subharmonic functions, for any $r_1 < r < r_2$, noticing that $\max_{|x|=r_2} \psi(x) \le 0$,

$$\max_{|x|=r} \psi(x) \le \frac{1}{\log \frac{r_2}{r_1}} \left(\max_{|x|=r_1} \psi(x) \log \frac{r_2}{r} + \max_{|x|=r_2} \psi(x) \log \frac{r}{r_1} \right)$$
$$\le \frac{1}{\log \frac{r_2}{r_1}} \max_{|x|=r_1} \psi(x) \log \frac{r_2}{r}.$$

By letting $r_2 \to \infty$, we get, for any $z \in \mathbb{C}$ with $0 < r_1 < |z|$, $\psi(z) \le \max_{|x|=r_1} \psi(x)$. Because ψ is upper semicontinuous, this implies that

$$\psi(z) \le \limsup_{r_1 \to 0} \max_{|x|=r_1} \psi(x) = \psi(0),$$

that is, ψ has a maximum at $z = 0$ and therefore is constant by the maximum principle for subharmonic functions. Thus $K \equiv 0$ by (3.12), so the surface must be a plane. □

The Bernstein theorem was extended by Heinz [41] in 1952 to get the following Gauss curvature estimate. Its proof is omitted here.

Theorem 3.3.3 ([41]). *Let M be a minimal graph over a disk of radius R. Then $|K(0)| \le c/R^2$, where c is a universal constant.*

The graph of a function $z = f(x, y)$ is a very particular surface. Such a surface passes a "vertical line test", that is, no two points on the surface will lie on a given vertical line. In other words, the image of its Gauss map is contained in either the upper hemisphere or the lower hemisphere. Therefore the image of its Gauss map always omits half of the unit sphere. For a general (not necessarily the graph of some function) complete nonplanaer minimal surface, Nirenberg took the point of view about the image of the Gauss map and made the following conjecture.

Nirenberg's conjecture. A complete simply connected minimal surface must be a plane if the normals
1. omit a neighborhood of some direction or
2. omit more than two directions (compare with the little Picard theorem).

R. Osserman in 1959–1965 showed that the Nirenberg conjecture 1 is true and 2 is false: The Gauss map to the classical Scherk surface omits four points. Xavier [88] in 1981 showed that the unit normals of a nonflat complete minimal surface in $\mathbb{R}^3$ cannot omit seven directions. Fujimoto [27] in 1988 solved the long-standing conjecture: The normals to a complete nonplanar minimal surface cannot omit more than four directions. The finite version (bound of the Gauss curvature) is also obtained. Below we present their proofs.

We first define the meaning of completeness. Let M be a surface. A continuous curve $\gamma(t)$, $0 \le t \le 1$, in M is said to be *divergent* in M if for each compact set G, there is t_0 such that $\gamma(t) \notin G$ for any $t \geq t_0$. We define the distance $d(p)(\leq +\infty)$ from a point $p \in M$ to the boundary as the greatest lower bound of the lengths of all continuous curves that are divergent in M. A surface M is said to be *complete* if $d(p) = \infty$ for all $p \in M$.

Theorem 3.3.4 (Osserman [56, 59]). *Let M be a complete minimal surface immersed in $\mathbb{R}^3$. If its Gauss map is not dense, then M is a plane.*

Proof. Let $\mathbf{x} = (x_1, x_2, x_3) : M \to \mathbb{R}^3$ be an immersion. Take a local isothermal coordinate (U, ϕ) for M, and let $\mathbf{x}(u, v) = (\mathbf{x} \circ \phi^{-1})(u, v)$. Let $\phi_k = \frac{\partial x_k}{\partial z}$ for $z = u + iv$. Then ϕ_k is holomorphic for $k = 1, 2, 3$. Let $f = \phi_1 - i\phi_2$ and $g = \frac{\phi_3}{\phi_1 - i\phi_2}$; then the induced metric is $ds^2 = |f|^2(1+|g|^2)^2|dz|^2$. As we know, g is the Gauss map. So, assuming that the Gauss map of M is not dense, we can assume that g omits ∞ and there exists $C > 0$ such that $|g| < C$. Thus $ds^2 = |f|^2(1 + |g|^2)^2|dz|^2 \leq (1 + C)|f|^2|dz|^2$, and hence the flat metric $d\bar{s}^2 := |f|^2|dz|^2$ is also complete. By Lemma 3.3.1, M with the flat metric $d\bar{s}^2 := |f|^2|dz|^2$ is isometric to the complex plane $\mathbb{C}$. Therefore g, which is bounded, is a holomorphic function on the whole complex plane. By Liouville's theorem, g is constant, and thus $K_{ds^2} \equiv 0$. Hence the surface must be a plane by Theorem 3.1.1. □

Theorem 3.3.5 (Fujimoto, [27]). *Let M be a complete minimal surface immersed in $\mathbb{R}^3$. If its Gauss map omits more than four values in S^2, then M lies in a plane.*

Proof. The proof is similar to the above, except that instead of using the fact $\int_0^{R_0} |dw| = a_0 R_0 < \infty$, we use the fact that $\int_0^1 \frac{dt}{t^p} < \infty$ for any $0 < p < 1$. We also need to use Corollary 2.3.9 to estimate $(1 + |g|^2)$ since in our case, $|g| < C$ is no longer true.

By lifting g to the universal cover of M, we can assume that M is simply connected. So M is biholomorphic to either $\mathbb{C}$ or the unit-disc. If $M = \mathbb{C}$, then by the little Picard theorem, g is constant. So the surface is flat. Hence we can assume that M is bioholomorphic to the unit disc. The induced metric, according to (3.8), is

$$ds^2 = |f|^2(1 + |g|^2)^2|dz|^2.$$

We assume that g omits $a_1, \ldots, a_q \in \mathbb{C} \cup \{\infty\}$ with $q \geq 5$ and $a_q = \infty$. We also assume that g is not constant. Take a positive η such that $q - 6 < q\eta < q - 4$. Similarly to the proof of Theorem 3.3.4, we seek a flat metric $d\bar{s}^2$ on M, which allows us to get a local isometry $w = \Phi$ (or change of coordinate by letting $w = \Phi(z)$) from the disk $D(R)$ to M and a divergent path in M, as stated in Lemma 3.3.1. As in the proof of Lemma 3.3.1, the flat metric (or Φ) can be decided through computing dw/dz (see (3.16)). Note that

$$\Phi^* ds = |(f \circ \Phi)(w)|(1 + |(g \circ \Phi)(w)|^2)\left|\frac{dz}{dw}\right||dw|.$$

To estimate $\Phi^* ds$, we want

$$\Phi^* ds = \left(\frac{|h'(w)|(1 + |h(w)|^2)^{1/p}}{\prod_{j=1}^{q-1} |h(w) - a_j|^{1-\eta}} \right)^p |dw|, \tag{3.13}$$

where $p = 2/(q - 2 - q\eta)$ and $h(w) = (g \circ \Phi)(w)$. Note that the purpose of doing so is to estimate $\Phi^* ds$ by using Corollary 2.3.9 in Chapter 2 for $h(w) = (g \circ \Phi)(w)$ on $D(R)$, that is,

$$\frac{|h'|(1 + |h|^2)^{1/p}}{\prod_{j=1}^{q-1} |h - a_j|^{1-\eta}} \le C \frac{R}{R^2 - |z|^2}. \tag{3.14}$$

Notice that (3.13) is the same as

$$|(f \circ \Phi)(w)| \left| \frac{dz}{dw} \right| = \left(\frac{|h'(w)|}{\prod_{j=1}^{q-1} |h(w) - a_j|^{1-\eta}} \right)^p. \tag{3.15}$$

Note that, by the chain rule, $h' = \frac{dg}{dz} \frac{dz}{dw}$, and thus (3.15) becomes

$$|(f \circ \Phi)(w)| \left| \frac{dz}{dw} \right| = \left(\frac{|\frac{dg}{dz}||\frac{dz}{dw}|}{\prod_{j=1}^{q-1} |(g \circ \Phi)(w) - a_j|^{1-\eta}} \right)^p,$$

that is,

$$\left| \frac{dw}{dz} \right|^{1-p} = \frac{|f(z)| \prod_{j=1}^{q-1} |g(z) - a_j|^{p(1-\eta)}}{|g'(z)|^p},$$

or

$$\left| \frac{dw}{dz} \right| = \frac{|f(z)|^{\frac{1}{1-p}} \prod_{j=1}^{q-1} |g(z) - a_j|^{\frac{p(1-\eta)}{1-p}}}{|g'(z)|^{\frac{p}{1-p}}}. \tag{3.16}$$

This suggests the flat metric we desire. To carry out the proof, we thus define the following new flat metric (with zero Gauss curvature):

$$d\sigma^2 := C^{-\frac{p}{1-p}} \frac{|f(z)|^{\frac{2}{1-p}} \prod_{j=1}^{q-1} |g(z) - a_j|^{\frac{2p(1-\eta)}{1-p}}}{|g'(z)|^{\frac{2p}{1-p}}} |dz|^2$$

on $M' = M - \{g' = 0\}$. With the flat metric $d\sigma^2$ on M', fix a point $p_0 \in M'$. By Lemma 3.3.1 there exists a local diffeomorphism Φ of a disk $D(R) = \{w \in \mathbb{C} : |w| < R\}$ $(0 < R \le \infty)$ onto an open neighborhood of p_0 with $\Phi(0) = p_0$ such that Φ is a local isometry. Since g is not constant, $R < \infty$ by the little Picard theorem. In addition, by Lemma 3.3.1 there exists a point a_0 with $|a_0| = 1$ such that the Φ-image Γ_{a_0} of the curve $L_{a_0} = \{w = a_0 t : 0 < t < R\}$ is divergent in $M' = M - \{g' = 0\}$. From the above discussion, with the choice of our flat metric, (3.13) holds, that is,

$$\Phi^* ds = \left(\frac{|h'(w)|(1 + |h(w)|^2)^{1/p}}{\prod_{j=1}^{q-1} |h(w) - a_j|^{1-\eta}} \right)^p |dw|.$$

We claim that the path Γ_{a_0} does not diverge to zeros of g'. Indeed, if Γ_{a_0} diverges to a point z_0 such that $g'(z_0) = 0$, then by taking a local coordinate on V, a neighborhood of z_0, with $\zeta(z_0) = 0$, we can write $d\sigma^2 = |\zeta|^{\frac{-2p}{1-p}} \beta |d\zeta|^2$ with $\beta > 0$ on V. Thus, because $p/(1-p) > 1$,

$$R = \int_{\Gamma_{a_0}} d\sigma \geq C \int_{\Gamma_{a_0} \cap V} \frac{1}{|\zeta|^{\frac{p}{1-p}}} |d\zeta| = \infty,$$

which gives a contradiction. Hence the path diverges to the boundary M. Denote by length(Γ_{a_0}) the length of the part Γ_{a_0} on M with respect to the induced metric ds^2. Then

$$\begin{aligned} \text{length}(\Gamma_{a_0}) &= \int_{\Gamma_{a_0}} ds = \int_{L_{a_0}} \Phi^* ds \\ &= \int_0^R \left(\frac{|h'(w)|(1 + |h(w)|^2)^{1/p}}{\prod_{j=1}^{q-1} |h(w) - a_j|^{1-\eta}} \right)^p |dw| \\ &\leq C' \int_0^R \frac{R^p}{(R^2 - t^2)^p} dt < \infty, \end{aligned}$$

where $p = 2/(q - 2 - q\eta) < 1$ when $q \geq 5$. This contradicts with the fact that the induced metric $ds^2 := |f|^2(1 + |g|^2)^2|dz|^2$ is complete. Therefore g must be constant, and thus the surface lies in a plane by Theorem 3.1.1. □

Actually, Fujimoto's five-point theorem can be reformulated to the following "complex analysis" version (see Theorem 5.2.1 for the related results):

Theorem 3.3.6 ([46]). *Let M be an open Riemann surface, and let g be a meromorphic function on M. Consider the conformal metric on M given by*

$$ds^2 = (1 + |g|^2)^m |\omega|^2,$$

where ω is a holomorphic 1-form, and $m \in \mathbb{Z}_{\geq 0}$. Assume that M is complete with respect to this metric. If g is nonconstant, then g can omit at most $m+2$ distinct values in $\mathbb{C} \cup \{\infty\}$.

The following is Xavier's result [88]. Although it is weaker than that of Fujimoto, his proof uses the integral results, especially the logarithmic derivative lemma in Nevanlinna theory, which we will prove in the next chapter (Proposition 4.1.6). His method allows us to study maps from balls in $\mathbb{C}^n$ into the complex projective spaces, which is an independent theory described in Chapter 6.

Theorem 3.3.7 ([88]). *Let M be a complete minimal surface immersed in $\mathbb{R}^3$. If its Gauss map omits more than seven values in S^2, then M must be a plane.*

The proof of Theorem 3.3.7 is based on the following result of Yau [90]. We first prove a lemma.

Lemma 3.3.8 ([90]). *Let M^n be a complete Riemannian manifold of dimension n, and let ω be a smooth integrable $(n-1)$-form defined on M^n. Then there exists a sequence of domains $\{B_i\}$ in M^n such that $B_i \subset B_{i+1}$, $M^n = \bigcup_i B_i$, and $\lim_{i\to\infty} \int_{B_i} d\omega = 0$.*

Proof. Let r be the distance function on M^n to a (fixed) point $p \in M^n$. Then r is a Lipschitz function on M^n. By approximating the function r (cf. [35]) we can find a nonnegative smooth function g_R such that

(1) For all but a finite number of $t < R$, $g_R^{-1}(t)$ is a compact regular hypersurface;

(2) $|dg_R| \le 3/2$ on $g_G^{-1}([0, R])$;

(3) $g_R^{-1}(t) \subset B(t+1)\backslash B(t-1)$ for $t \le R$, where $B(R)$ is a ball centered at p with radius R.

On the other hand, Theorem 3.2.22 of [22] shows that

$$\int_{g_R^{-1}([0,R])} |dg_R||\omega| = \int_0^R \left(\int_{g_R^{-1}(t)} |\omega| \right) dt.$$

It follows from (2) that

$$\int_0^R \left(\int_{g_R^{-1}(t)} |\omega| \right) dt \le \frac{3}{2} \int_M |\omega|.$$

Therefore, for some $R/2 \le t_R \le R$, where $g_R^{-1}(t_R)$ is a compact regular hypersurface,

$$\int_{g_R^{-1}(t_R)} |\omega| \le \frac{3}{R} \int_M |\omega|.$$

By Stokes' theorem,

$$\left| \int_{g_R^{-1}([0,t_R])} d\omega \right| \le \int_{g_R^{-1}(t_R)} |\omega| \le \frac{3}{R} \int_M |\omega|.$$

This, together with property (3) of g_R, we get

$$M^n = \bigcup_i g_R^{-1}([0, t_i])$$

and

$$\lim_{i\to\infty} \int\limits_{g_R^{-1}([0,t_i])} d\omega = 0. \qquad \square$$

Theorem 3.3.9 (Yau [90]). *Let M be a complete Riemannian manifold of infinite volume. Let u be a nonnegative smooth function satisfying $\triangle \log u \geq 0$ almost everywhere. Then $\int_M u^p dA = \infty$ for any $p > 0$ unless u is a constant function.*

Proof. We only prove the case where $p = 1$ (the general case is essentially the same since $\triangle \log u^p = p \triangle \log u \geq 0$).

For every $\epsilon > 0$, let $v_\epsilon = (u + \epsilon)^{1/2}$. Then

$$\triangle \log v_\epsilon \geq \frac{\epsilon |du|^2}{2u(u+\epsilon)^2} \geq 0,$$

that is,

$$v_\epsilon \triangle v_\epsilon \geq |dv_\epsilon|^2. \tag{3.17}$$

Let r be the distance function on M to a (fixed) point $p \in M$. For arbitrary R_1, R_2 such that $0 < R_1 < R_2$, let

$$\omega = \phi\left(\frac{r + R_2 - 2R_1}{R_2 - R_1}\right),$$

where ϕ is a smooth function such that $0 \leq \phi \leq 1$, $\phi(t) = 1$ for $t \leq 1$, and $\phi(t) = 0$ for $t \geq 2$. Then ω is a continuous Lipschitz function, $0 \leq \omega \leq 1$, and $\omega(x) = 1$ for $x \in B(R_1)$ and $\omega(x) = 0$ for $x \in M \backslash B(R_2)$. Also, $|d\omega| \leq C/(R_2 - R_1)$, where $C > 0$ is a constant. By the Stokes theorem,

$$\int\limits_{B(R_2)} dv_\epsilon \wedge *d(\omega^2 v_\epsilon) = -\int\limits_{B(R_2)} \omega^2 v_\epsilon \triangle v_\epsilon. \tag{3.18}$$

By (3.17) and (3.18), we have

$$\int\limits_{B(R_2)} \omega^2 |dv_\epsilon|^2 \leq -\int\limits_{B(R_2)} 2v_\epsilon dv_\epsilon \wedge *\omega d\omega - \int\limits_{B(R_2)} \omega^2 |dv_\epsilon|^2.$$

Hence

$$2\int\limits_{B(R_2)} \omega^2 |dv_\epsilon|^2 \leq \int\limits_{B(R_2)} \omega^2 |dv_\epsilon|^2 + \int\limits_{B(R_2)} v_\epsilon^2 |d\omega|^2.$$

Since $|d\omega| \leq C/(R_2 - R_1)$, we see that

$$\frac{1}{4}\int_{B(R_2)} \frac{\omega^2|du|^2}{u+\epsilon} = \int_{B(R_2)} \omega^2|dv_\epsilon|^2 \le \frac{C^2}{(R_2-R_1)^2}\int_{B(R_2)} v_\epsilon^2.$$

By taking $\epsilon \to 0$ we get

$$\frac{1}{4}\int_{B(R_2)} \frac{\omega^2|du|^2}{u} \le \frac{C^2}{(R_2-R_1)^2}\int_{B(R_2)} u.$$

We now prove the theorem by contradiction. Assume that $\int_M u dA < \infty$ and u is not a constant function. By taking $R_2 = 2R_1$ and letting $R_1 \to \infty$ the above inequality becomes

$$\int_M \frac{|du|^2}{u} < \infty.$$

By the Schwarz inequality we get

$$\left(\int_M |du|\right)^2 \le \int_M \frac{|du|^2}{u}\int_M u < \infty.$$

Let $u_\epsilon = u + \epsilon$. Then

$$\triangle \log u_\epsilon = \frac{\epsilon}{u(u+\epsilon)^2}|du|^2.$$

Since

$$\int_M |d\log u_\epsilon| = \int_M \frac{|du|}{u+\epsilon} \le \int_M \frac{|du|}{\epsilon} < \infty,$$

if follows from Lemma 3.3.8 that there exists a sequence of domains $\{B_i\}$ such that

$$0 = \lim_{i\to\infty}\int_{B_i} \triangle \log u_\epsilon = \int_M \frac{\epsilon}{u(u+\epsilon)^2}|du|^2 \ge 0$$

for all $\epsilon > 0$. Hence

$$\int_M \frac{|du|^2}{u(u+\epsilon)^2} = 0.$$

This implies that u is constant, a contradiction. Hence $\int_M u dA = \infty$. □

Lemma 3.3.10. *Let g be a nowhere zero holomorphic function on the unit disk $\triangle(1)$. Assume that for $0 < r < 1$,*

$$T_g(r) \le C\log\frac{1}{1-r},$$

where $T_g(r)$ is Nevanlinna's characteristic function defined in Chapter 4 (see (4.8)). Then there exists a positive constant K such that for $0 < r < 1$,

$$\frac{1}{2\pi}\int_0^{2\pi}\left|\frac{g'}{g}(re^{i\theta})\right|d\theta \le \frac{K}{1-r}\log\frac{1}{1-r}.$$

Proof. By the assumption, $\log|g|$ is harmonic on $\triangle(1)$. For a given z in $D(R)$ with $R < 1$, we consider the linear transformation $L(w) = \frac{R^2(z-w)}{R^2-\bar{z}w}$, which sends z to zero and satisfies $|L(w)| = R$ if $|w| = R$. Let $G(w) = \log|g(L(w))|$. Applying Theorem 2.2.1 (mean value theorem) for harmonic functions to $G(w)$, we have

$$\log|g(z)| = G(0) = \int_0^{2\pi}\log G(\mathrm{Re}^{i\theta})\frac{d\theta}{2\pi} = \int_{|w|=R} G(w)\frac{dw}{2\pi i w}.$$

We let $\zeta = L(w)$. Then

$$w = L^{-1}(\zeta) = \frac{R^2(z-\zeta)}{R^2-\bar{z}\zeta}.$$

So, for $|\zeta| = R$,

$$\begin{aligned}\frac{dw}{2\pi i w} &= \frac{1}{2\pi i}\left(\frac{-1}{z-\zeta}+\frac{\bar{z}}{R^2-\bar{z}\zeta}\right)d\zeta = \left(\frac{-1}{z-\zeta}+\frac{\bar{z}}{\zeta\bar{\zeta}-\bar{z}\zeta}\right)\frac{d\zeta}{2\pi i}\\ &= \left(\frac{-\zeta}{z-\zeta}+\frac{\bar{z}}{\bar{\zeta}-\bar{z}}\right)\frac{d\zeta}{2\pi i\zeta} = \frac{R^2-|z|^2}{|\zeta-z|^2}\frac{d\zeta}{2\pi i\zeta}.\end{aligned}$$

Note that when $|w| = R$, $|\zeta| = R$, and $\frac{d\zeta}{i\zeta} = d\theta$, we get the following Poisson formula:

$$\log|g(z)| = \int_0^{2\pi}\log|g(\mathrm{Re}^{i\theta})|\frac{R^2-|z|^2}{|\,\mathrm{Re}^{i\theta}-z|^2}\frac{d\theta}{2\pi}. \tag{3.19}$$

As

$$\mathrm{Re}\left(\frac{\zeta+z}{\zeta-z}\right) = \frac{|\zeta|^2-|z|^2}{|\zeta-z|^2},$$

using the fact that

$$\log|f(z)| = \frac{1}{2}[\log f(z) + \log\bar{f}(z)]$$

and $(\log\bar{f}(z))' = 0$, and thus

$$\frac{f'(z)}{f(z)} = (\log f(z))' = 2(\log|f(z)|)'$$

for any nowhere zero holomorphic function f, by differentiating both sides of (3.19), we get

$$\left(\frac{g'}{g}\right)(z) = 2\int_0^{2\pi} \log|g(\mathrm{Re}^{i\phi})|\frac{\mathrm{Re}^{i\phi}}{(\mathrm{Re}^{i\phi}-z)^2}\frac{d\phi}{2\pi}. \tag{3.20}$$

On the other hand, by Corollary 4.1.2,

$$\int_0^{2\pi} \frac{d\theta}{|\mathrm{Re}^{i\phi}-re^{i\theta}|^2} = \int_0^{2\pi} \frac{d\theta}{|R-re^{i\theta}|^2} = \frac{2\pi}{R^2-r^2}.$$

Since $|\log|x|| = \log^+ x + \log^+(1/x)$ for $x > 0$, by the first main theorem (see Chapter 4) we have

$$\int_0^{2\pi} |\log|g(\mathrm{Re}^{i\phi})||\frac{d\phi}{2\pi} = m_g(R,\infty) + m_g(R,0) \le 2T_g(R) + O(1).$$

By the assumption $T_g(R) \le C\log\frac{1}{1-R}$ we conclude that

$$\frac{1}{2\pi}\int_0^{2\pi}\left|\frac{g'}{g}(re^{i\theta})\right|d\theta \le \frac{C'}{(R-r)}\log\frac{1}{1-R}$$

for some constant $C' > 0$. By taking $R = (1+r)/2$ we obtain the desired inequality. □

Proof of Theorem 3.3.7. By lifting g to the universal cover of M we can assume that M is simply connected. So M is biholomorphic to either $\mathbb{C}$ or the unit disc. If $M = \mathbb{C}$, then by Picard's theorem, g is constant. So the surface is flat. Without loss of generality, we assume that $M = \triangle(1)$, the unit disc. By applying Theorem 4.1.10 (see Chapter 4) the little Picard theorem still holds for $g : \triangle(1) \to \mathbb{P}^1 = \mathbb{C} \cup \{\infty\}$ if $\limsup_{r\to 1}\frac{T_g(r)}{\log(1-r)} = \infty$. Hence we still can get that g is constant (so the surface is flat) under the assumption that $\limsup_{r\to 1}\frac{T_g(r)}{\log(1-r)} = \infty$. It remains to prove the case that

$$T_g(r) \le C\log\frac{1}{1-r}.$$

Assume that g omits ∞ and 0. Then, by Lemma 3.3.10,

$$\frac{1}{2\pi}\int_0^{2\pi}\left|\frac{g'}{g}(re^{i\theta})\right|d\theta \le \frac{K}{1-r}\log\frac{1}{1-r} \tag{3.21}$$

for some constant $K > 0$. We claim that on $\triangle(1)$ we have the following inequality:

$$\frac{|g'|^{5/6}(1+|g|^2)^2}{\prod_{j=1}^{6}|g-a_j|^{5/6}} \le C\sum_{i=1}^{6}\frac{|g'|^{5/6}}{|g-a_j|^{5/6}} = C\sum_{i=1}^{6}\left|\frac{(g-a_j)'}{(g-a_j)}\right|^{5/6}.$$

Indeed, let $\delta = \min_{i\neq j}\{|a_i - a_j|, 1\}$. For z, let i_0 be the index among $\{1,\dots,6\}$ such that $|g(z) - a_{i_0}| \le |g(z) - a_j|$ for $1 \le j \le 6$. Then, for $j \ne i_0$, by the triangle inequality

$$\delta \le |a_{i_0} - a_j| \le |g(z) - a_{i_0}| + |g(z) - a_j| \le 2|g(z) - a_j|,$$

and hence $|g(z) - a_j| > \delta/2$. Thus

$$\frac{|g'(z)|^{5/6}(1+|g(z)|^2)^2}{\prod_{j=1}^{6}|g(z)-a_j|^{5/6}} \le K_1\frac{|g'(z)|^{5/6}(1+|g(z)|^2)^2}{|g(z)-a_{i_0}|^{5/6}}$$

for some constant $K_1 > 0$. If $|g(z) - a_{i_0}| \le 1/3$, then g is bounded, so the claim holds. If $|g(z) - a_{i_0}| > 1/3$ (so $|g(z) - a_j| > 1/3$ for all $1 \le j \le 6$), then $\frac{(1+|g(z)|^2)^2}{\prod_{j=1, j\ne i_0}^{6}|g(z)-a_j|^{5/6}}$ is bounded, and hence the claim still holds. This proves the claim.

Thus by (3.21) we get

$$\int_0^{2\pi}\left(\frac{|g'|^{5/6}(1+|g|^2)^2}{\prod_{j=1}^{6}|g-a_j|^{5/6}}\right)(re^{i\theta})d\theta = C\sum_{i=1}^{6}\int_0^{2\pi}\left|\frac{(g-a_j)'}{(g-a_j)}(re^{i\theta})\right|^{5/6}d\theta$$
$$\le C'\sum_{i=1}^{6}\left(\int_0^{2\pi}\left|\frac{(g-a_j)'}{(g-a_j)}(re^{i\theta})\right|d\theta\right)^{5/6} \le \frac{C''}{(1-r)^{5/6}}(\log\frac{1}{1-r})^{5/6}$$

for some positive constants C, C', C''. Note that the induced metric on $\triangle(1)$ is $ds^2 = |f|^2(1+|g|^2)^2|dz|^2$. Note that since g omits ∞, f is holomorphic without zeros. Take $u := \frac{|g'|}{|f|^{12/5}\prod_{j=1}^{6}|g-a_j|}$. Then

$$\int_{\triangle(1)} u^{5/6}dA = \int_{\triangle(1)}\frac{|g'|^{5/6}(1+|g|^2)^2}{\prod_{j=1}^{6}|g-a_j|^{5/6}}dudv$$
$$= \int_0^1 rdr\left(\int_0^{2\pi}\left(\frac{|g'|^{5/6}(1+|g|^2)^2}{\prod_{j=1}^{6}|g-a_j|^{5/6}}\right)(re^{i\theta})d\theta\right)$$
$$\le K\int_0^r \frac{r}{(1-r)^{5/6}}\left(\log\frac{1}{(1-r)}\right)^{5/6}dr < \infty.$$

So $\int_{\triangle(1)} u^p dA < \infty$ for $p = 5/6$. Note that u satisfies $\triangle \log u = 0$ almost everywhere, and also the minimal surface M has infinite volume (every complete simply connected surface with nonpositive curvature has infinite volume). Hence, by Theorem 3.3.9, $\int_{\triangle(1)} u^p dA = \infty$, a contradiction. □

3.4 Minimal surfaces with finite total curvatures

In this section, we study the Gauss map of minimal surfaces of finite curvature. See [34, 44]. In particular, we prove Osserman's result that the Gauss map of a nonflat complete minimal surface in $\mathbb{R}^3$ with finite total curvature can omit at most three values.

We begin with the following result.

Theorem 3.4.1. *Let M be a complete Riemanian manifold of dimension 2 whose Gauss curvature $K \leq 0$. Assume that*

$$\int_M |K| dA < \infty.$$

Then there exist a compact 2-manifold $\tilde{M}$, a finite number of points $p_1, \ldots, p_k$ on $\tilde{M}$, and an isometry between M and $\tilde{M} - \{p_1, \ldots, p_k\}$.

This is a deep result, and its proof is not presented in this book. See Theorem 9.1 in Osserman's book [62] for the proof.

As we have seen earlier, the Gauss curvature of minimal surfaces is always nonpositive, so complete minimal surfaces of $\mathbb{R}^3$ with finite total curvature satisfy the hypothesis of the above theorem. We claim that after identification of M with $\tilde{M} - \{p_1, \ldots, p_k\}$, the Gauss map $g : M \to \mathbb{P}^1(\mathbb{C})$, which is holomorphic, extends to a meromorphic function on $\tilde{M}$. Indeed, if any of the points p_j were an essential singularity of g, then by the big Picard theorem in complex analysis, g would assume all values of $\mathbb{P}^1(\mathbb{C})$ infinitely many times, with at most two exceptions. Since the Gauss map g is the unit normal, this means that the unit normal $\mathbf{n}$ would assume all values of $S^2(1)$ (the unit sphere) infinitely many times, with at most two exceptions. This would imply that the total curvature of M would be infinite (by the geometric meaning of the Gauss curvature; see Theorem 1.2.3), which is contrary to our assumption. Hence g has at most a pole at p_j for $1 \leq j \leq k$. Thus it can be extended to a meromorphic function on $\tilde{M}$. This proves the claim.

Recall that under the isothermal coordinates (u, v) and the immersion $\mathbf{x} : M \to \mathbb{R}^3$, if $\phi := \frac{\partial \mathbf{x}}{\partial z}$, where $z = u + iv$, then $\alpha_k = \phi_i dz$, $1 \leq k \leq 3$, are global holomorphic 1-forms on M, where $\phi = (\phi_1, \phi_2, \phi_3)$. Also recall that the induced metric is $ds^2 = |f|^2(1+|g|^2)^2|dz|^2$, where $f = \phi_1 - i\phi_2$. We have that the form $\omega := f dz$ extends as a meromorphic form to $\tilde{M}$. Indeed, by changing coordinates in $\mathbb{R}^3$ we may assume that g has no poles at the points $p_1, \ldots, p_k$. Since the metric of M is complete, we have

$$\lim_{z \to p_j} |f(z)| = \infty.$$

Hence f extends as a meromorphic with at most poles at $p_j, 1 \le j \le k$, on $\tilde{M}$. Since $\phi_j, 1 \le j \le 3$, can be expressed through f and g, we see that α_k, $1 \le k \le 3$, extend to meromorphic forms on $\tilde{M}$.

Proposition 3.4.2. *Let M be a complete minimal surfaces in $\mathbb{R}^3$ with finite total curvature. Then*

$$\int_M |K|dA = 4\pi \deg(g),$$

where $g : \tilde{M} \to \mathbb{P}^1(\mathbb{C})$ is the holomorphic extension of $g : M \to \mathbb{P}^1(\mathbb{C})$.

Proof. The induced metric is $ds^2 = 2\|\phi\|^2|dz|^2 = |f|^2(1+|g|^2)^2|dz|^2$, where $\phi = \phi_1, \phi_2, \phi_2$. Its associated metric form is $\psi := \|\phi\|^2 \frac{\sqrt{-1}}{2\pi} dz \wedge d\bar{z} = |f|^2(1+|g|^2)^2 \frac{\sqrt{-1}}{4\pi} dz \wedge d\bar{z}$. Hence, using $\mathrm{Ric}(\psi) = -K\psi$ and (2.12), we have

$$\begin{aligned}
\int_M |K|dA &= -\int_{\tilde{M}} KdA = -\int_{\tilde{M}} K2\|\phi\|^2 dudv = -2\int_{\tilde{M}} K\|\phi\|^2 du \wedge dv \\
&= -2\int_{\tilde{M}} K\|\phi\|^2 \frac{\sqrt{-1}}{2} dz \wedge d\bar{z} = -2\pi \int_{\tilde{M}} K\psi = 2\pi \int_M \mathrm{Ric}(\psi) \\
&= 2\pi \int_{\tilde{M}} dd^c \log(|f|^2(1+|g|^2)^2) \\
&= 2\pi \int_{\tilde{M}} dd^c \log(1+|g|^2)^2 = 4\pi \deg(g).
\end{aligned} \tag{3.22}$$

□

The induced metric $ds^2 = 2\sum_{i=1}^3 |\phi_i(z)|^2|dz|^2$ extends to a pseudo-metric on $\tilde{M}$. Write $ds^2 = |z|^{2\nu_j} h^2|dz|^2$ locally near p_j with $h(0) \ne 0$ and a local coordinate z with $z(p_j) = 0$. We prove that

$$\nu_j \le -2. \tag{3.23}$$

To do so, we prove that the form $\alpha = (\alpha_1, \alpha_2, \alpha_3)$, where $\alpha_j = \phi_j dz$, has a pole of order $m_j \ge 2$ at each p_j. Indeed, take a local coordinate z around p_j with $z(p_j) = 0$. At $z = 0$, α_i has at most a pole of order m_{ij} at p_j. Since $\tilde{M} - \{p_1, \dots, p_k\}$ is complete in the induced metric given by $ds^2 = \sum_{i=1}^3 |\phi_i(z)|^2|dz|^2$ near p_j, we have

$$\lim_{z \to p_j} \sum_{i=1}^3 |\phi_i(z)|^2 = \infty.$$

Thus $m_j = \max\{m_{1j}, m_{2j}, m_{3j}\} \ge 1$. Observe that m_j is exactly the order of the pole of α at p_j. Assume that $m_1 = 1$. Then

$$\phi_i(z) = \frac{c_{-1}^i}{z} + \sum_{n=0}^{\infty} c_n^i z^n, \quad i = 1, 2, 3.$$

Since

$$x_i = \operatorname{Re} \int \phi_i(z)dz = \operatorname{Re}\left(c_i \log z + \sum_{n=0}^{\infty} \frac{c_n^i}{n+1} z^{n+1} \right) + x_i(z_0),$$

we know that the term $\operatorname{Re}(c_{-1}^i \log z)$ must be single valued because all other terms are single valued, which implies that c_i must be real. On the other hand,

$$0 = \sum_{i=1}^{3} \phi_i^2 = \frac{(c_{-1}^1)^2 + (c_{-1}^2)^2 + (c_{-1}^3)^2}{z^2} + \frac{d_{-1}}{z} + \sum_{n=0}^{\infty} d_n z^n,$$

and it follows that $\sum_{i=1}^{3}(c_{-1}^i)^2 = 0$, so $c_i = 0$ for $i = 1, 2, 3$. However, this is impossible since M is complete with respect to ds^2. Therefore we have that $\alpha_j = \phi_j dz$ has a pole of order $m_j \geq 2$ at each p_j. From the definition of ds^2 we see that (3.23) holds.

By applying Theorem 2.4.1 (Gauss–Bonnet for pseudo-metric) to $\psi = |f|^2(1 + |g|^2)^2 \frac{\sqrt{-1}}{4\pi} dz \wedge d\bar{z}$, the pseudo-metric form associated with ds^2, we get

$$-\int_{\tilde{M}} \operatorname{Ric}(\psi) = \chi(\tilde{M}) + \sum_{j=1}^{k} \nu_j.$$

By (3.22),

$$-\int_{\tilde{M}} \operatorname{Ric}(\psi) = -\int_{\tilde{M}} dd^c \log(|f|^2(1 + |g|^2)^2) = -\int_{\tilde{M}} dd^c \log(1 + |g|^2)^2 = -2\deg(g).$$

On the other hand, from (3.23) we get $\nu_j \leq -2$ for $j = 1, 2, \ldots, k$. Hence

$$-2\deg(g) = \chi(\tilde{M}) + \sum_{j=1}^{k} \nu_j \leq \chi(\tilde{M}) - 2k. \tag{3.24}$$

Theorem 3.4.3 ([61]). *Let* $\mathbf{x} : M \to \mathbb{R}^3$ *be a nonflat complete minimal surface of finite total curvature. Then its Gauss map can omit at most three points (i. e., three directions on the unit sphere).*

Proof. Since the minimal surface $\mathbf{x} : M \to \mathbb{R}^3$ has finite total curvature, as we discussed above, M is conformally equivalent to a compact surface $\tilde{M}$ punctured at a finite number of points $P = p_1, \ldots, p_k$, and the Gauss map g can be extended to a meromorphic function on $\tilde{M}$. Hence $g : \tilde{M} \to \mathbb{P}^1(\mathbb{C})$ is a holomorphic map. Assume that g omits $a_1, \ldots, a_q \in \mathbb{P}^1(\mathbb{C})$. By Theorem 2.4.3 with $E = f^{-1}\{a_1, \ldots, a_q\}$ we have

$$(q-2)d \le |E| + 2(g_{\tilde{M}} - 1) \le k + 2(g_{\tilde{M}} - 1),$$

where $d = \deg(g)$, and $g_{\tilde{M}}$ is the genus of $\tilde{M}$. On the other hand, from (3.24), $g_{\tilde{M}} - 1 \le d - k$. Therefore

$$(q-2)d \le k + 2(g_{\tilde{M}} - 1) \le k + 2(d-k) = 2d - k,$$

or

$$k \le (4-q)d.$$

Since $k > 0$, we conclude that $q < 4$. □

4 The Nevanlinna theory

In this chapter, we introduce the theory of meromorphic functions developed by Nevanlinna [50] in 1925, as well its extension to holomorphic curves in the complex projective space. For references, see [76, 9, 10, 19, 34, 50, 51, 52, 80]. Other related works and recent developments include [8, 15, 17, 21, 23, 36, 37, 69, 70, 74, 75, 78, 81, 82].

The fundamental tool of this subject is the measurement of the growth of f. There are two different ways of measuring an entire function f: its rate of growth—its maximum modulus on the disc of radius r (viewed as a function of r) and the maximum number of times the value in the image is taken on this disc. The insight is that these two rates of growth are essentially the same, the former being roughly the exponential of the latter. For meromorphic functions, the maximum modulus no longer exists if the function has poles. Nevanlinna [50] found the right substitute for the maximum modulus. He introduced the characteristic function $T_f(r)$ to measure the growth of a meromorphic function f. Starting from the Poisson–Jensen formula, he was able to derive a more subtle growth estimate for meromorphic functions, which he called the *second main theorem*. It gives a quantitative extension of the classical Picard theorem for meromorphic functions.

The theory of holomorphic curves, which extends the theory of meromorphic functions, is mainly due to Cartan [10] and Ahlfors [1, 2]. It is the heart part of the so-called Nevanlinna theory. In the mid-1960s, Chern and Osserman [20] initiated a project by using the theory of holomorphic curves to study value distribution properties of the Gauss maps of the complete minimal surfaces in $\mathbb{R}^m$.

4.1 Nevanlinna theory of meromorphic functions

A function f defined on a domain in $D \subset \mathbb{C}$ is called a *meromorphic function* on D if it is holomorphic on D except the poles. In this section, we introduce the Nevanlinna theory of meromorphic functions on $D(R) = \{z \mid |z| < R\}$ with $0 < R \leq \infty$. The reference book is [40]. The theory of meromorphic functions gives a quantitative formulation of the little Picard theorem. The main tool is the integration result of the so-called *logarithmic derivative lemma*. We begin with the following Poisson–Jensen formula.

Theorem 4.1.1 (Poisson–Jensen formula). *Let $f \not\equiv 0$ be a meromorphic function on the closed disc $\overline{D}(R)$, where $D(R) = \{z \mid |z| < R\}$, $R < \infty$. Let $a_1, \ldots, a_p$ denote the zeros of f in $D(R)$, counting multiplicities, and let $b_1, \ldots, b_q$ denote the poles of f in $D(R)$, also counting multiplicities. Then, for any z in $|z| < R$ that is not a zero or pole, we have*

$$\log|f(z)| = \int_0^{2\pi} \frac{R^2 - |z|^2}{|\mathrm{Re}^{i\theta} - z|^2} \log|f(\mathrm{Re}^{i\theta})| \frac{d\theta}{2\pi}$$

https://doi.org/10.1515/9783110989557-004

$$- \sum_{i=1}^{p} \log \left| \frac{R^2 - \bar{a}_i z}{R(z - a_i)} \right| + \sum_{j=1}^{q} \log \left| \frac{R^2 - \bar{b}_j z}{R(z - b_j)} \right|.$$

Proof. Note that it suffices to prove the theorem when f has no zeros or poles on the circle $|z| = R$. Otherwise, we consider the function $f(\rho z)$ and let $\rho \to 1$.

We first consider the case where f is analytic and has no zeros in the closed disc $|z| \leq R$. In this case, $\log |f|$ is harmonic. For a given z in $D(R)$, we consider the linear transformation $L(w) = \frac{R^2(z-w)}{R^2 - \bar{z}w}$. The function L sends z to zero and satisfies $|L(w)| = R$ if $|w| = R$. Let $F(w) = \log |f(L(w))|$. Applying Theorem 2.2.1 (mean value theorem) to harmonic functions to $F(w)$, we have

$$\log|f(z)| = F(0) = \int_0^{2\pi} F(\mathrm{Re}^{i\theta}) \frac{d\theta}{2\pi} = \int_{|w|=R} F(w) \frac{dw}{2\pi i w}. \tag{4.1}$$

Let $\zeta = L(w)$. Then

$$w = L^{-1}(\zeta) = \frac{R^2(z - \zeta)}{R^2 - \bar{z}\zeta}.$$

So, for $|\zeta| = R$,

$$\begin{aligned} \frac{dw}{2\pi i w} &= \frac{1}{2\pi i}\left(\frac{-1}{z - \zeta} + \frac{\bar{z}}{R^2 - \bar{z}\zeta} \right) d\zeta = \left(\frac{-1}{z - \zeta} + \frac{\bar{z}}{\bar{\zeta}\zeta - \bar{z}\zeta} \right) \frac{d\zeta}{2\pi i} \\ &= \left(\frac{-\zeta}{z - \zeta} + \frac{\bar{z}}{\bar{\zeta} - \bar{z}} \right) \frac{d\zeta}{2\pi i \zeta} = \frac{R^2 - |z|^2}{|\zeta - z|^2} \frac{d\zeta}{2\pi i \zeta}. \end{aligned} \tag{4.2}$$

Note that when $|w| = R$, $|\zeta| = R$, and $\frac{d\zeta}{i\zeta} = d\theta$, so by combining (4.1) and (4.2)

$$\log|f(z)| = \int_0^{2\pi} \log|f(\mathrm{Re}^{i\theta})| \frac{R^2 - |z|^2}{|\,\mathrm{Re}^{i\theta} - z|^2} \frac{d\theta}{2\pi}.$$

Thus

$$\log|f(z)| = \int_0^{2\pi} \log|f(\mathrm{Re}^{i\theta})| \frac{R^2 - |z|^2}{|\,\mathrm{Re}^{i\theta} - z|^2} \frac{d\theta}{2\pi}. \tag{4.3}$$

Thus the theorem holds in this case.

For the general case, we consider the function

$$g(z) = f(z) \frac{\prod_{\mu=1}^{p} \frac{R^2 - \bar{a}_\mu z}{R(z - a_\mu)}}{\prod_{\nu=1}^{q} \frac{R^2 - \bar{b}_\nu z}{R(z - b_\nu)}}.$$

Then g has no zeros or poles in $|z| \leq R$. Note that when $|z| = R$, $|g(z)| = |f(z)|$. Applying (4.3) to g yields the theorem. □

Applying the above theorem with $f(z) \equiv e$, we have the following:

Corollary 4.1.2. *Let $|z| < R$. Then*

$$\int_0^{2\pi} \frac{1}{|\,\mathrm{Re}^{i\theta} - z|^2} \frac{d\theta}{2\pi} = \frac{1}{R^2 - |z|^2}.$$

We now define Nevanlinna functions. Let f be a meromorphic function on $D(R_0)$, where $0 < R_0 \leq \infty$, and let $r < R_0$. Denote the number of poles of f on the open disc $D(r)$ by $n_f(r, \infty)$, counting multiplicities. Define the *counting function*

$$N_f(r, \infty) = n_f(0, \infty) \log r + \int_0^r [n_f(t, \infty) - n_f(0, \infty)] \frac{dt}{t},$$

where $n_f(0, \infty)$ is the multiplicity if f has a pole at $z = 0$. For each complex number a, we define

$$N_f(r, a) = N_{1/(f-a)}(r, \infty). \tag{4.4}$$

We note that $N_f(r, a)$ measures how many times f takes the value a in $|z| < r$. By the definition of the Lebesgue–Stieltjes integral it follows that

$$N_f(r, 0) = (\mathrm{ord}_0^+ f) \log r + \sum_{|z|<r, z \neq 0} (\mathrm{ord}_z^+ f) \log \left| \frac{r}{z} \right|, \tag{4.5}$$

where $\mathrm{ord}_z^+ f = \max\{0, \mathrm{ord}_z f\}$ is just the multiplicity of the zero of f at z. With this notation, from the Poisson–Jensen formula we have

Corollary 4.1.3 (Jensen's formula). *Let $f \not\equiv 0$ be meromorphic on $\overline{D}(r)$. Then*

$$\log |c_f| = \int_0^{2\pi} \log |f(re^{i\theta})| \frac{d\theta}{2\pi} - \sum_{z \in D(r), z \neq 0} (\mathrm{ord}_z f) \log \left| \frac{r}{z} \right| - (\mathrm{ord}_0 f) \log r,$$

or, equivalently,

$$\log |c_f| = \int_0^{2\pi} \log |f(re^{i\theta})| \frac{d\theta}{2\pi} + N_f(r, \infty) - N_f(r, 0),$$

where c_f is the leading nonzero coefficient in the Taylor expansion of f around 0.

The Nevanlinna *proximity function* is defined as

$$m_f(r,\infty) = \int_0^{2\pi} \log^+ |f(re^{i\theta})| \frac{d\theta}{2\pi}, \tag{4.6}$$

where $\log^+ x = \max\{0, \log x\}$. For any complex number a, we define

$$m_f(r,a) = m_{1/(f-a)}(r,\infty). \tag{4.7}$$

Note that $m_f(r,a)$ measures how close f is, on average, to a on the circle of radius r. Finally, the *Nevanlinna characteristic function* of f is defined by

$$T_f(r) = m_f(r,\infty) + N_f(r,\infty). \tag{4.8}$$

It measures the growth of f. For example, $T_f(r) = O(1)$ if and only if f is constant, and $T_f(r) = O(\log r)$ if and only if f is a rational function (see [40]).

The characteristic function T, the proximity function m, and the counting function N are the three *Nevanlinna functions*. Nevanlinna theory can be described as the study of how the growth of these three functions is interrelated. The first main theorem is a reformulation of Corollary 4.1.3.

Theorem 4.1.4 (First main theorem). *Let $f \not\equiv 0$ be meromorphic on $\overline{D}(R_0)$, $0 < R_0 \le \infty$. Then, for any $0 < r < R_0$,*

(i) $T_f(r) = m_f(r,0) + N_f(r,0) + \log |c_f|$;

(ii) *For a complex number a,*

$$\left|T_f(r) - m_f(r,a) - N_f(r,a)\right| \le \left|\log |c_{1/(f-a)}|\right| + \log^+ |a| + \log 2,$$

where $c_{1/(f-a)}$ is the leading nonzero coefficient in the Taylor expansion of $1/(f-a)$ around 0.

Proof. (i) is derived directly from Corollary 4.1.3. To prove (ii), applying Corollary 4.1.3 to $1/(f-a)$ yields

$$\log |c_{1/(f-a)}| = \int_0^{2\pi} \log \frac{1}{|f(re^{i\theta}) - a|} \frac{d\theta}{2\pi} + N_{1/(f-a)}(r,\infty) - N_{1/(f-a)}(r,0).$$

Since $\log x = \log^+ x - \log^+(1/x)$,

$$\log |c_{1/(f-a)}| = \int_0^{2\pi} \log^+ \frac{1}{|f(re^{i\theta}) - a|} \frac{d\theta}{2\pi} - \int_0^{2\pi} \log^+ |f(re^{i\theta}) - a| \frac{d\theta}{2\pi} + N_f(r,a) - N_f(r,\infty).$$

Thus

$$\int_0^{2\pi} \log^+ |f(re^{i\theta}) - a| \frac{d\theta}{2\pi} = -N_f(r,\infty) + m_f(r,a) + N_f(r,a) - \log |c_{1/(f-a)}|.$$

Note that if x and y are positive real numbers, then

$$\log^+(x+y) \le \log^+ 2\max\{x,y\} \le \log^+ x + \log^+ y + \log 2.$$

So

$$\big|\log^+ |x-y| - \log^+ |x|\big| \le \log^+ |y| + \log 2.$$

Thus

$$\big|T_f(r) - m_f(r,a) - N_f(r,a) + \log |c_{1/(f-a)}|\big| \le \log^+ |a| + \log 2. \qquad \square$$

To prove the *logarithmic derivative lemma*, we first establish the following result.

Lemma 4.1.5. *Let $r > 0$ and $0 < p < 1$. For every $a \in \mathbb{C}$, we have*

$$\int_0^{2\pi} \frac{r^p}{|re^{i\theta} - a|^p} \frac{d\theta}{2\pi} \le \frac{(2-p)}{2(1-p)}.$$

Proof. Without no loss of generality, we may assume that a is a positive real number. Then for $|\theta| \le \pi/2$, we have

$$|re^{i\theta} - a| \ge r|\sin\theta| \ge \frac{2}{\pi} r|\theta|,$$

and for $\pi/2 < |\theta| \le \pi$, we have $|re^{i\theta} - a| \ge r$. Therefore

$$\begin{aligned}\int_0^{2\pi} \frac{r^p}{|re^{i\theta} - a|^p} d\theta &\le 2\int_0^{\pi/2} \left(\frac{\pi}{2\theta}\right)^p d\theta + 2\int_{\pi/2}^{\pi} d\theta \\ &\le \frac{2^{1-p}\pi^p}{1-p}\left(\frac{\pi}{2}\right)^{1-p} + \pi = \frac{\pi(2-p)}{1-p}.\end{aligned} \qquad \square$$

Proposition 4.1.6. *Let f be a non-constant meromorphic function in $D(R_0)$ $(0 < R_0 \le \infty)$. Let p and p' be real numbers such that $0 < p < p' < 1$. Let $0 < r_0 < R_0$. Then there exists a positive constant K such that for $r_0 < r < \rho < R_0$,*

$$\int_0^{2\pi} \left| r\left(\frac{f'}{f}\right)(re^{i\theta})\right|^p \frac{d\theta}{2\pi} \le K\left(\frac{\rho}{\rho - r} T_f(\rho)\right)^{p'}. \tag{4.9}$$

Proof. We start from the Poisson–Jensen formula: for $0 < r < s < R_0$,

$$\log|f(z)| = \int_0^{2\pi} \frac{s^2 - |z|^2}{|se^{i\phi} - z|^2} \log|f(se^{i\phi})| \frac{d\phi}{2\pi} - \sum_{|a_\mu|<s} \log\left|\frac{s^2 - \bar{a}_\mu z}{s(z - a_\mu)}\right| + \sum_{|b_\nu|<s} \log\left|\frac{s^2 - \bar{b}_\nu z}{s(z - b_\nu)}\right|. \tag{4.10}$$

Since

$$\log|f(z)| = \frac{1}{2}[\log f(z) + \log \bar{f}(z)]$$

and $(\log \bar{f}(z))' = 0$,

$$\frac{f'(z)}{f(z)} = (\log f(z))' = 2(\log|f(z)|)'.$$

Differentiating the both sides of (4.10) and noticing that

$$\mathrm{Re}\left(\frac{\zeta + z}{\zeta - z}\right) = \frac{|\zeta|^2 - |z|^2}{|\zeta - z|^2},$$

we get

$$\frac{f'(z)}{f(z)} = \int_0^{2\pi} \frac{2se^{i\phi}}{(se^{i\phi} - z)^2} \log|f(se^{i\phi})| \frac{d\phi}{2\pi} - \sum_{|a_\mu|<s} \left(\frac{1}{a_\mu - z} - \frac{\bar{a}_\mu}{s^2 - \bar{a}_\mu z}\right) + \sum_{|b_\nu|<s} \left(\frac{1}{b_\nu - z} - \frac{\bar{b}_\nu}{s^2 - \bar{b}_\nu z}\right).$$

Hence

$$\left|z\left(\frac{f'}{f}\right)(z)\right| \le 2s|z| \int_0^{2\pi} \frac{|\log|f(se^{i\phi})||}{|se^{i\phi} - z|^2} \frac{d\phi}{2\pi} + \sum_{|a_\mu|<s} \left(\frac{|z|}{|a_\mu - z|} + \frac{|a_\mu||z|}{|s^2 - \bar{a}_\mu z|}\right) + \sum_{|b_\nu|<s} \left(\frac{|z|}{|b_\nu - z|} + \frac{|b_\nu z|}{|s^2 - \bar{b}_\nu z|}\right). \tag{4.11}$$

Therefore, for $z = re^{i\theta}$ and $\tilde{p} = p/p'$,

$$\left|r\left(\frac{f'}{f}\right)(re^{i\theta})\right|^{\tilde{p}} \le \left(2sr \int_0^{2\pi} \frac{|\log|f(se^{i\phi})||}{|se^{i\phi} - re^{i\theta}|^2} \frac{d\phi}{2\pi}\right)^{\tilde{p}}$$

$$+ \sum_{|a_\mu|<s} \left\{ \left| \frac{r}{a_\mu - re^{i\theta}} \right|^{\tilde{p}} + \left| \frac{r}{(s^2/\bar{a}_\mu) - re^{i\theta}} \right|^{\tilde{p}} \right\}$$

$$+ \sum_{|b_\nu|<s} \left\{ \left| \frac{r}{b_\nu - re^{i\theta}} \right|^{\tilde{p}} + \left| \frac{r}{(s^2/\bar{b}_\nu) - re^{i\theta}} \right|^{\tilde{p}} \right\}.$$

In the following, K_i $(i = 1, 2, \dots)$ denote some suitable constants. Integrating each term with respect to θ and using Lemma 4.1.5, we have

$$\begin{aligned}
&\int_0^{2\pi} \left| r\left(\frac{f'}{f}\right)(re^{i\theta}) \right|^{\tilde{p}} \frac{d\theta}{2\pi} \\
&\quad \le \int_0^{2\pi} \frac{d\theta}{2\pi} \left(2sr \int_0^{2\pi} \frac{|\log|f(se^{i\phi})||}{|se^{i\phi} - re^{i\theta}|^2} \frac{d\phi}{2\pi} \right)^{\tilde{p}} + K_1(n_f(s,\infty) + n_f(s,0)) \\
&\quad \le \left(\int_0^{2\pi} \frac{d\theta}{2\pi} \left(\int_0^{2\pi} 2sr \frac{|\log|f(se^{i\phi})||}{|se^{i\phi} - re^{i\theta}|^2} \frac{d\phi}{2\pi} \right) \right)^{\tilde{p}} + K_1(n_f(s,\infty) + n_f(s,0)).
\end{aligned}$$

On the other hand, by Corollary 4.1.2

$$\int_0^{2\pi} \frac{d\theta}{|se^{i\phi} - re^{i\theta}|^2} \le \frac{2\pi}{(s^2 - r^2)} \le \frac{\pi}{r(s-r)},$$

since $s + r \ge 2r$. Hence

$$\begin{aligned}
\int_0^{2\pi} \left| r\left(\frac{f'}{f}\right)(re^{i\theta}) \right|^{\tilde{p}} \frac{d\theta}{2\pi} \le \left(\frac{s}{s-r} \int_0^{2\pi} |\log|f(se^{i\phi})|| \frac{d\phi}{2\pi} \right)^{\tilde{p}} \\
+ K_1(n_f(s,\infty) + n_f(s,0)).
\end{aligned} \tag{4.12}$$

By the first main theorem

$$\int_0^{2\pi} |\log|f(se^{i\phi})|| \frac{d\phi}{2\pi} = m_f(s,\infty) + m_f(s,0) \le 2T_f(s) + K_2.$$

Also, taking $s = (\rho + r)/2$, we have

$$N_f(\rho, 0) + |\operatorname{ord}_0 f| \log^+ \frac{1}{r_0} \ge \int_s^\rho n_f(t,0) \frac{dt}{t} \ge n_f(s,0) \int_s^\rho \frac{dt}{\rho} = \frac{\rho - s}{\rho} n_f(s,0).$$

So by the first main theorem

$$\begin{aligned} n_f(s,0) &\leq \frac{\rho}{\rho-s}\left(N_f(\rho,0)+|\operatorname{ord}_0 f|\log^+\frac{1}{r_0}\right) \\ &\leq \frac{\rho}{\rho-s}\left(T_f(\rho)+|\operatorname{ord}_0 f|\log^+\frac{1}{r_0}\right) \\ &\leq \frac{2\rho}{\rho-r}(T_f(\rho)+K_4), \end{aligned} \tag{4.13}$$

and, similarly,

$$n_f(s,\infty) \leq \frac{2\rho}{\rho-r}(T_f(\rho)+K_5).$$

Hence

$$\int_0^{2\pi}\left|r\left(\frac{f'}{f}\right)(re^{i\theta})\right|^{\tilde{p}}\frac{d\theta}{2\pi} \leq K_6\left(\frac{\rho}{\rho-r}T_f(\rho)\right)^{\tilde{p}}+K_7\frac{\rho}{\rho-r}T_f(\rho).$$

By Hölder's inequality with $\tilde{p}=p/p'$

$$\begin{aligned} \int_0^{2\pi}\left|r\left(\frac{f'}{f}\right)(re^{i\theta})\right|^{p}\frac{d\theta}{2\pi} &\leq \left(\int_0^{2\pi}\left|r\left(\frac{f'}{f}\right)(re^{i\theta})\right|^{\tilde{p}}\frac{d\theta}{2\pi}\right)^{p'} \\ &\leq K_8\left(\frac{\rho}{\rho-r}T_f(\rho)\right)^{\tilde{p}p'}+K_9\left(\frac{\rho}{\rho-r}T_f(\rho)\right)^{p'} \leq K_{10}\left(\frac{\rho}{\rho-r}T_f(\rho)\right)^{p'}. \end{aligned}$$

This proves our proposition. □

Before we derive the lemma of the logarithmic derivative, we need the following lemma.

Lemma 4.1.7 (Borel's growth lemma).

(a) *Let $F(r)$ be a positive nondecreasing continuous function on $[r_0,\infty)$ with $r_0 \geq e$ such that $F(r) \geq e$ on $[r_0,\infty)$. Then for every $\epsilon > 0$, there exists a closed set $E \subset [r_0,\infty)$ (called the "exceptional set") of finite Lebesgue measure such that if we set $\rho = r + 1/\log^{1+\epsilon} F(r)$ for all $r \geq r_0$ not in E, then we have*

$$\log F(\rho) \leq \log F(r) + 1 \tag{4.14}$$

and

$$\log^+ \frac{\rho}{r(\rho-r)} \leq (1+\epsilon)\log^+\log F(r) + \log 2. \tag{4.15}$$

(b) *Let $F(r)$ be a positive nondecreasing continuous function defined on $[r_0,R_0)$ with $R_0 < \infty$ and $F(r) \geq e$ on $[r_0,R_0)$. Then there exists a closed set $E \subset [r_0,R_0)$ with $\int_E (dr/(R_0-r)) \leq 2$ such that if we set $\rho = r + (R_0-r)/eF(r)$, then*

$$\log F(\rho) \le \log F(r) + 1 \tag{4.16}$$

for all $r \in [0, R_0)$ *not in* E.

Proof. (a) Let

$$E = \left\{ r \in [r_0, \infty) : F\left(r + \frac{1}{\log^{1+\epsilon} F(r)} \right) \ge eF(r) \right\}.$$

We may assume that E is nonempty; otherwise, the lemma is trivial. We claim that E is of finite Lebesgue measure.

Let r_1 be the smallest $r \in E$ such that $r \ge r_0$. Now assume that we have found numbers $r_1, \ldots, r_n, s_1, \ldots, s_{n-1}$. We describe here how to inductively extend this set, and we continue this process as long as possible. If there is no number s such that $F(s) \ge eF(r_n)$, then we stop. Otherwise, by the continuity of F there exists s such that $F(s) = eF(r_n)$. Let s_n be the smallest such s. If there is $r \in E$ such that $r \ge s_n$, then let r_{n+1} be the smallest such r. Otherwise, we stop.

For each pair r_j, s_j, clearly, $s_j > r_j$, and since $r_j \in E$,

$$F\left(r_j + \frac{1}{\log^{1+\epsilon} F(r_j)} \right) \ge eF(r_j) = F(s_j).$$

Since F is nondecreasing, this implies

$$r_j + \frac{1}{\log^{1+\epsilon} F(r_j)} \ge s_j,$$

and so

$$s_j - r_j \le \frac{1}{\log^{1+\epsilon} F(r_j)}. \tag{4.17}$$

Moreover, $F(r_{j+1}) \ge F(s_j) = eF(r_j)$ since $s_{j+1} \ge s_j$. Hence

$$F(r_{n+1}) \ge eF(r_n) \ge e^2 F(r_{n-1}) \ge \cdots \ge e^n F(r_1) \ge e^{n+1}. \tag{4.18}$$

It follows that either we can only find finitely many r_n or the sequence r_n goes to the infinity as n goes to the infinity. Since the set E is contained in the union of $[r_n, s_n]$, if we can only find finitely many r_n, then E is of finite Lebesgue measure. Now consider the case where n goes to ∞. Let $m(E)$ be the Lebesgue measure of E. Then

$$m(E) \le \sum_{n=1}^{\infty} (s_n - r_n).$$

By (4.17) and (4.18)

$$\sum_{n=1}^{\infty}(s_n - r_n) \leq \sum_{n=1}^{\infty} \frac{1}{\log^{1+\epsilon} F(r_n)} \leq \sum_{n=1}^{\infty} \frac{1}{n^{1+\epsilon}} < +\infty.$$

Thus the claim is proved.

To verify (4.14), let $r \geq r_0$, where r is not contained in E. Then by the construction of E

$$F(\rho) = F\left(r + \frac{1}{\log^{1+\epsilon} F(r)}\right) \leq eF(r).$$

Thus $\log F(\rho) \leq \log F(r) + 1$. So (4.14) holds. Finally, we verify (4.15). We have

$$\frac{\rho}{r(\rho - r)} = \frac{1}{\rho - r} + \frac{1}{r} \leq \log^{1+\epsilon} F(r) + 1 \leq 2\log^{1+\epsilon} F(r).$$

Hence

$$\log^{+} \frac{\rho}{r(\rho - r)} \leq (1 + \epsilon)\log^{+} \log F(r) + \log 2.$$

This proves (a).

For the proof of (b), see Lemma 2.4 (ii) in Hayman's book [40]. □

Proposition 4.1.8. *Suppose that a positive real-valued function $T(r)$ on $[0,1)$ satisfies the condition*

$$T(r) \leq \frac{1}{(1-r)^p}$$

for a set $E_0 = \bigcup_{\nu=1}^{\infty}[r_\nu, r'_\nu](r'_{\nu-1} < r_\nu \leq r'_\nu)$ such that $\int_{E_0} \frac{dr}{1-r} < \infty$ and a positive number p. Then there exists a positive constant K such that $T(r) \leq \frac{K}{(1-r)^p}$ for every $r \in [0,1)$.

Proof. We may replace $T(r)$ by $T^*(r) = \sup_{r' \leq r} T(r')$. So we may assume that $T(r)$ is an increasing function. By the assumption we see that

$$\sum_{\nu} \int_{r_\nu}^{r'_\nu} \frac{dr}{1-r} = \sum_{\nu} \log \frac{1-r_\nu}{1-r'_\nu} = K_0 < \infty.$$

For r such that $r_\nu \leq r \leq r'_\nu$, we have

$$\begin{aligned} T(r) &\leq \frac{1}{(1-r'_\nu)^p} = \frac{1}{(1-r_\nu)^p}\left(\frac{1-r_\nu}{1-r'_\nu}\right)^p \\ &\leq \frac{e^{pK_0}}{(1-r_\nu)^p} \leq \frac{e^{pK_0}}{(1-r)^p}. \end{aligned}$$

By taking $K = e^{pK_0}$ we derive our result. □

Theorem 4.1.9 (Lemma on the logarithmic derivative). *Let f be a nonconstant meromorphic function on $D(R_0)$ $(0 < R_0 \le \infty)$. Assume that $T_f(r_0) \ge e$ for some $r_0 \in (0, R_0)$. Then for all $\epsilon > 0$,*

$$\frac{1}{2\pi}\int_0^{2\pi} \log^+\left|\frac{f'}{f}(re^{i\theta})\right| d\theta \le S(r) + \epsilon \log^+ r$$

for $r_0 < r < R_0$, where $S(r)$ is evaluated as follows:

(i) *In the case $R_0 < \infty$,*

$$S(r) \le K\left(\log^+ \frac{1}{R-r} + \log^+ T_f(r)\right)$$

for every $r \in [0, R_0)$ excluding a set E such that $\int_E \frac{dt}{R_0 - t} < \infty$.

(ii) *In the case $R_0 = \infty$,*

$$S(r) \le (1+\epsilon)\log^+ T_f(r) + (1+\epsilon)^2 \log^+ \log T_f(r) + O(1)$$

for every $r \in [0, \infty)$ excluding a set E such that $\int_E dt < \infty$.

Proof. Using the concavity of $\log^+$, we pull the $\log^+$ outside the integral and use Proposition 4.1.6 to get for $r_0 < r < \rho < R_0$ such that $0 < p < p' < 1$,

$$\begin{aligned} m_{f'/f}(r,\infty) &= \frac{1}{p}\int_0^{2\pi} \log^+\left|\frac{f'(re^{i\theta})}{f(re^{i\theta})}\right|^p \frac{d\theta}{2\pi} \le \frac{1}{p}\log^+ \int_0^{2\pi}\left|\frac{f'(re^{i\theta})}{f(re^{i\theta})}\right|^p \frac{d\theta}{2\pi} \\ &\le \frac{p'}{p}\left(\log^+ \frac{\rho}{r(\rho-r)} + \log^+ T_f(\rho)\right) + \frac{p'-p}{p}\log^+ r + O(1) \\ &\le (1+\epsilon)\left(\log^+ \frac{\rho}{r(\rho-r)} + \log^+ T_f(\rho)\right) + \epsilon \log^+ r + O(1) \\ &\le S(r) + \epsilon\log^+ r + O(1) \end{aligned} \tag{4.19}$$

by taking p' such that $\frac{p'}{p} \le 1+\epsilon$, where

$$S(r) := (1+\epsilon)\left(\log^+ \frac{\rho}{r(\rho-r)} + \log^+ T_f(\rho)\right) + O(1).$$

Assume that $R_0 = \infty$. Take $\rho = r + \frac{1}{\log^{1+\epsilon} T_f(r)}$. Applying Lemma 4.1.7 with $F(r) = T_f(r)$, for $r \ge r_0$ not contained in E such that $\int_E dt < \infty$, we have

$$\log T_f(\rho) \le \log T_f(r) + 1 \tag{4.20}$$

and

$$\log^+ \frac{\rho}{r(\rho - r)} \leq (1+\epsilon)\log^+ \log T_f(r) + \log 2. \tag{4.21}$$

Hence

$$S(r) \leq_{\text{exc}} (1+\epsilon)\log^+ T_f(r) + (1+\epsilon)^2 \log^+ \log T_f(r) + O(1),$$

where $\leq_{\text{exc}}$ means the inequality for all $r \notin E$. In the case $R_0 < \infty$, by Lemma 4.1.7 we get

$$T_f\left(r + \frac{R_0 - r}{eT_f(r)}\right) \leq eT_f(r)$$

for every $r \in [0, R_0)$ excluding a set E such that $\int_E \frac{dt}{R_0 - t} < \infty$. Thus

$$S(r) \leq K\left(\log^+ \frac{1}{R - r} + \log^+ T_f(r)\right)$$

for every $r \in [0, R_0)$ excluding a set E such that $\int_E \frac{dt}{R_0 - t} < \infty$. □

Before we state the second main theorem, we introduce the *ramification term* $N_{\text{ram},f}(r)$ of f defined by

$$N_{\text{ram},f}(r) = N_{f'}(r, 0) + 2N_f(r, \infty) - N_{f'}(r, \infty).$$

Note that $N_{\text{ram},f}(r) \geq 0$.

Theorem 4.1.10 (The second main theorem). *Let $a_1, \ldots, a_q$ be a set of distinct complex numbers. Let f be a nonconstant meromorphic function on $D(R_0)$ with $0 < R_0 \leq \infty$. Then for all $\epsilon > 0$ and $0 < r < R_0$, we have*

$$\begin{aligned}(q-1)T_f(r) &+ N_{\text{ram},f}(r) \\ &\leq \sum_{j=1}^{q} N_f(r, a_j) + N_f(r, \infty) + \epsilon \log^+ r + S(r),\end{aligned}$$

where $S(r)$ is evaluated as follows:

(i) *In the case $R_0 < \infty$,*

$$S(r) \leq K\left(\log^+ \frac{1}{R_0 - r} + \log^+ T_f(r)\right)$$

for every $r \in [0, R)$ excluding a set E such that $\int_E \frac{dt}{R_0 - t} < \infty$.

(ii) *In the case $R_0 = \infty$,*

$$S(r) \leq (1+\epsilon)\log^+ T_f(r) + (1+\epsilon)^2 \log^+ \log T_f(r) + O(1)$$

for every $r \in [0, \infty)$ *excluding a set E such that* $\int_E dt < \infty$.

Proof. Let $\delta = \min_{i \neq j}\{|a_i - a_j|, 1\}$. For each z with $f(z) \neq \infty$ and $f(z) \neq a_j$ for $1 \leq j \leq q$, let j_0 be the index among $\{1, 2, \ldots, q\}$ such that

$$|f(z) - a_{j_0}| \leq |f(z) - a_j| \quad \text{for all } 1 \leq j \leq q.$$

Then for $j \neq j_0$, by the triangle inequality, $|f(z) - a_j| \geq \delta/2$. Thus, for $j \neq j_0$,

$$\begin{aligned}\log^+|f(z)| &\leq \log^+|f(z) - a_j| + \log^+|a_j| + \log 2\\ &\leq \log|f(z) - a_j| + \log^+\frac{2}{\delta} + \log^+|a_j| + \log 2.\end{aligned}$$

Therefore

$$(q-1)\log^+|f(z)| \leq \sum_{j \neq j_0} \log|f(z) - a_j| + \sum_{j=1}^{q} \log^+|a_j| + (q-1)\left(\log^+\frac{2}{\delta} + \log 2\right).$$

Now

$$\begin{aligned}\sum_{j \neq j_0} \log|f(z) - a_j| &= \sum_{j=1}^{q} \log|f(z) - a_j| - \log|f'(z)| + \log\frac{|f'(z)|}{|f(z) - a_{j_0}|}\\ &\leq \sum_{j=1}^{q} \log|f(z) - a_j| - \log|f'(z)| + \log\left(\sum_{j=1}^{q} \frac{|f'(z)|}{|f(z) - a_j|}\right).\end{aligned}$$

Thus

$$\begin{aligned}(q-1)\log^+|f(z)| \leq{}& \sum_{j=1}^{q} \log|f(z) - a_j| - \log|f'(z)| + \log\left(\sum_{j=1}^{q} \frac{|f'(z)|}{|f(z) - a_j|}\right)\\ &+ \sum_{j=1}^{q} \log^+|a_j| + (q-1)\left(\log^+\frac{2}{\delta} + \log 2\right).\end{aligned}$$

Now set $z = re^{i\theta}$ and integrate with respect to θ to get

$$\begin{aligned}(q-1)m_f(r, \infty) &= (q-1)\int_0^{2\pi} \log^+|f(re^{i\theta})|\frac{d\theta}{2\pi}\\ &\leq \sum_{j=1}^{q} \int_0^{2\pi} \log|f(re^{i\theta}) - a_j|\frac{d\theta}{2\pi} - \int_0^{2\pi} \log|f'(re^{i\theta})|\frac{d\theta}{2\pi}\\ &\quad + \int_0^{2\pi} \log\left(\sum_{j=1}^{q} \frac{|f'(re^{i\theta})|}{|f(re^{i\theta}) - a_j|}\right)\frac{d\theta}{2\pi} + O(1).\end{aligned}$$

From Corollary 4.1.3 we have

$$\int_0^{2\pi} \log|f(re^{i\theta}) - a_j| \frac{d\theta}{2\pi} = N_f(r, a_j) - N_f(r, \infty) + \log|c_{f-a_j}|$$

and

$$\int_0^{2\pi} \log|f'(re^{i\theta})| \frac{d\theta}{2\pi} = N_{f'}(r, 0) - N_{f'}(r, \infty) + \log|c_{f'}|.$$

Thus the above inequality becomes

$$\begin{aligned}
&(q-1)m_f(r,\infty) - \sum_{j=1}^{q} N_f(r, a_j) + qN_f(r,\infty) + N_{f'}(r,0) - N_{f'}(r,\infty) \\
&\quad \le \int_0^{2\pi} \log\left(\sum_{j=1}^{q} \frac{|f'(re^{i\theta})|}{|f(re^{i\theta}) - a_j|} \right) \frac{d\theta}{2\pi} + \sum_{j=1}^{q} \log^+ |a_j| \\
&\quad\quad + (q-1)\left(\log^+ \frac{2}{\delta} + \log 2 \right) + \sum_{j=1}^{q} \log|c_{f-a_j}| - \log|c_{f'}|.
\end{aligned}$$

However, by the first main theorem and the definition of $N_{\mathrm{ram},f}(r)$ the left-hand side of the above inequality is

$$(q-1)T_f(r) - \sum_{j=1}^{q} N_f(r, a_j) - N_f(r,\infty) + N_{\mathrm{ram},f}(r).$$

To complete the proof, we still need to estimate

$$\int_0^{2\pi} \log\left(\sum_{j=1}^{q} \frac{|f'(re^{i\theta})|}{|f(re^{i\theta}) - a_j|} \right) \frac{d\theta}{2\pi}.$$

Let α be a real number between 0 and 1. Then by Proposition 4.1.6 and the inequality $\frac{\alpha'}{\alpha} \le 1 + \epsilon$,

$$\begin{aligned}
&\int_0^{2\pi} \log\left(\sum_{j=1}^{q} \frac{|f'(re^{i\theta})|}{|f(re^{i\theta}) - a_j|} \right) \frac{d\theta}{2\pi} \\
&\quad = \frac{1}{\alpha} \int_0^{2\pi} \log\left(\sum_{j=1}^{q} \frac{|f'(re^{i\theta})|}{|f(re^{i\theta}) - a_j|} \right)^{\alpha} \frac{d\theta}{2\pi} \\
&\quad \le \frac{1}{\alpha} \int_0^{2\pi} \log\left(\sum_{j=1}^{q} \left| \frac{f'(re^{i\theta})}{f(re^{i\theta}) - a_j} \right|^{\alpha} \right) \frac{d\theta}{2\pi} \le \frac{1}{\alpha} \log\left(\sum_{j=1}^{q} \int_0^{2\pi} \left| \frac{f'(re^{i\theta})}{f(re^{i\theta}) - a_j} \right|^{\alpha} \frac{d\theta}{2\pi} \right)
\end{aligned}$$

$$\leq \frac{\alpha'}{\alpha}\left(\log^+\left(\frac{\rho}{r(\rho-r)}\right) + \log^+ T_f(\rho)\right) + \frac{\alpha'-\alpha}{\alpha}\log^+ r + O(1)$$
$$\leq (1+\epsilon)\left(\log^+\left(\frac{\rho}{r(\rho-r)}\right) + \log^+ T_f(\rho)\right) + \epsilon\log^+ r + O(1),$$

where in the last inequality, we used the concavity of the logarithm and the inequality $(\sum d_j)^\alpha \leq \sum d_j^\alpha$ for positive real numbers d_j and $0 < \alpha < 1$. Thus

$$(q-1)T_f(r) - \sum_{j=1}^{q} N_f(r, a_j) - N_f(r, \infty) + N_{\mathrm{ram},f}(r) \leq \epsilon \log r + S(r),$$

where

$$S(r) = (1+\epsilon)\left(\log^+\left(\frac{\rho}{r(\rho-r)}\right) + \log^+ T_f(\rho)\right) + O(1).$$

The estimate of $S(r)$ can be carried out as in the previous theorem. □

Define the *truncated counting function* as

$$N_f^{(1)}(r, 0) = \min\{\mathrm{ord}_0^+ f, 1\}\log r + \sum_{z\in D(r), z\neq 0} \min\{\mathrm{ord}_z^+ f, 1\}\log\left|\frac{r}{z}\right|. \tag{4.22}$$

Theorem 4.1.10 can be restated as the following corollary.

Corollary 4.1.11. *Let $a_1, \ldots, a_q$ be a set of distinct complex numbers. Let f be a nonconstant meromorphic function on $D(R_0)$ with $0 < R_0 \leq \infty$. Then for any $\epsilon > 0$, we have the inequality*

$$(q-1)T_f(r) \leq \sum_{j=1}^{q} N_f^{(1)}(r, a_j) + N_f^{(1)}(r, \infty) + \epsilon \log^+ r + S(r)$$

for $0 < r < R_0$, where $S(r)$ is evaluated as in Theorem 4.1.10.

Proof. It is easy to check that

$$\sum_{j=1}^{q} N_f(r, a_j) + N_f(r, \infty) - N_{\mathrm{ram},f}(r) \leq \sum_{j=1}^{q} N_f^{(1)}(r, a_j) + N_f^{(1)}(r, \infty).$$

Combining this inequality and Theorem 4.1.10 concludes the corollary. □

Corollary 4.1.12 (Picard's theorem). *If a meromorphic function f on $\mathbb{C}$ omits three distinct points $a_1, a_2, a_3 \in \mathbb{C} \cup \{\infty\}$, then f is constant.*

Proof. Assume that f is not constant. Applying Theorem 4.1.10, we have

$$\sum_{j=1}^{3} m_f(r, a_j) \leq_{\text{exc}} 2T_f(r) + (1+\epsilon)\log^+ T_f(r) + (1+\epsilon)^2 \log^+ \log T_f(r) + \epsilon \log^+ r + O(1),$$

where $\leq_{\text{exc}}$ means the inequality for all $r \in (0,\infty)\backslash E$ with $\int_E dt < \infty$. However, since f omits a_j, $N_f(r, a_j) = 0$ for $1 \leq j \leq 3$. So $m_f(r, a_j) = T_f(r) + O(1)$. Thus

$$3T_f(r) \leq_{\text{exc}} 2T_f(r) + \log^+ T_f(r) + (1+\epsilon)\log^+ \log T_f(r) + \epsilon \log^+ r + O(1).$$

This is a contradiction. □

4.2 Cartan's second main theorem for holomorphic curves

To introduce the theory, we reformulate $T_f(r)$, $m_f(r,a)$, and $N_f(r,a)$, which appeared in the previous section. Let $\mathbb{P}^1(\mathbb{C})$ be the complex projective space of dimension 1, that is, $\mathbb{P}^1(\mathbb{C}) = \mathbb{C}^2 - \{0\}/\sim$, where $(a_1, a_2) \sim (b_1, b_2)$ if and only if $(a_1, a_2) = \lambda(b_1, b_2)$ for some $\lambda \in \mathbb{C}$. We denote by $[a_1 : a_2]$ the equivalence class of (a_1, a_2). The space $\mathbb{P}^1(\mathbb{C})$ is naturally identical to $\mathbb{C} \cup \{\infty\}$ by the following map

$$[a:1] \mapsto a \quad \text{for } a \in \mathbb{C} \quad \text{and} \quad [1:0] \mapsto \{\infty\}. \tag{4.23}$$

In this way, any meromorphic function f on $D(R)$, $0 < R \leq \infty$, determines a holomorphic map $f : D(R) \to \mathbb{P}^1(\mathbb{C})$ with $f(z) = [1:0]$ if z is a pole of f and $f(z) = [f(z):1]$ otherwise. We now reformulate the definitions. On $\mathbb{P}^1(\mathbb{C})$ (or $\mathbb{C} \cup \{\infty\}$), there is a natural distance called the *chordal distance* defined for all $z_1, z_2 \in \mathbb{C} \cup \{\infty\}$ as

$$\|z_1, z_2\|^2 = \begin{cases} \frac{|z_1 - z_2|^2}{(1+|z_1|^2)(1+|z_2|^2)} & \text{if } z_2 \neq \infty, \\ \frac{1}{(1+|z_1|^2)} & \text{if } z_2 = \infty. \end{cases} \tag{4.24}$$

Let $a \in \mathbb{C} \cup \{\infty\}$, and let $m_f(r,a)$ be the proximity function defined in the previous section. Using the inequality $\log^+ x \leq \log(1+x) \leq \log^+ x + \log 2$ for every $x > 0$, we can easily get, for $0 < r < R$,

$$m_f(r,a) = \int_0^{2\pi} \log \frac{1}{\|f(re^{i\theta}), a\|} \frac{d\theta}{2\pi} + O(1). \tag{4.25}$$

The counting function $N_f(r,a)$ stays the same. Let $f = f_1/f_2$, where f_1, f_2 are entire functions without common factors. By Corollary 4.1.3

$$N_f(r,\infty) = \int_0^{2\pi} \log|f_2(re^{i\theta})|\frac{d\theta}{2\pi} + O(1).$$

So the characteristic function is

$$\begin{aligned} T_f(r) &= m_f(r,\infty) + N_f(r,\infty) \\ &= \int_0^{2\pi} \frac{1}{2}\log(1+|f(re^{i\theta})|^2)\frac{d\theta}{2\pi} + \int_0^{2\pi} \log|f_2(re^{i\theta})|\frac{d\theta}{2\pi} \\ &= \int_0^{2\pi} \log(|f_1(re^{i\theta})|^2 + |f_2(re^{i\theta})|^2)^{1/2}\frac{d\theta}{2\pi}. \end{aligned} \tag{4.26}$$

We now extend the theory to holomorphic curves in $\mathbb{P}^n(\mathbb{C})$. Recall that the n-dimensional complex projective space is $\mathbb{P}^n(\mathbb{C}) = \mathbb{C}^{n+1} - \{0\}/\sim$, where $(a_0, a_1, \ldots, a_n) \sim (b_0, b_1, \ldots, b_n)$ if and only if $(a_0, \ldots, a_n) = \lambda(b_0, \ldots, b_n)$ for some $\lambda \in \mathbb{C}$. We denote by $[a_0 : \cdots : a_n]$ the equivalence class of $(a_0, \ldots, a_n)$. Let $f : D(R_0) \to \mathbb{P}^n(\mathbb{C})$ be a holomorphic map, where $0 < R_0 \leq \infty$. Denote $\mathbf{f} = (f_0, \ldots, f_n)$, which is called a *reduced representation* of f if $\mathbb{P}(\mathbf{f}) = f$ and $f_0, \ldots, f_n$ are entire functions without common zeros. Let $\|\mathbf{f}\|$ (or simply $\|f\|$) be $\|\mathbf{f}\| = \max_{0\leq i\leq n} |f_i|$. The *Cartan characteristic function* $T_f(r)$ of f is defined by

$$T_f(r) = \int_0^{2\pi} \log\|\mathbf{f}(re^{i\theta})\|\frac{d\theta}{2\pi} - \log\|\mathbf{f}(0)\|,\ 0 < r < R_0. \tag{4.27}$$

Note that (see [76]) up to a bounded term,

$$T_f(r) = \int_0^r \frac{dt}{t} \int_{Dt} f^*\omega_{FS},$$

where $\omega_{FS} = dd^c \log\|w\|^2$ is the Fubini–Study form on $\mathbb{P}^n(\mathbb{C})$. A hyperplane in $\mathbb{P}^n(\mathbb{C})$ is given by

$$H = \left\{[x_0 : \cdots : x_n] \in \mathbb{P}^n(\mathbb{C}) \mid \sum_{i=0}^n a_i x_i = 0\right\},$$

where $a_i \in \mathbb{C}, 0 \leq i \leq n$. Denote by $\mathbf{a} = (a_0, \ldots, a_n)$ the nonzero vector associated with H. The *Weil function* $\lambda_H(f)$ *of* f *with respect to the hyperplane* H is defined by

$$\lambda_H(f)(z) = \log\frac{\|\mathbf{f}(z)\|\|\mathbf{a}\|}{|\langle\mathbf{f}(z),\mathbf{a}\rangle|}, \tag{4.28}$$

where $\langle \mathbf{f}, \mathbf{a} \rangle$ is the inner product on $\mathbb{C}^{n+1}$. The *proximity function* $m_f(r, H)$ *of* f *with respect to* H is defined as

$$m_f(r,H) = \int_0^{2\pi} \lambda_H(f)(re^{i\theta}) \frac{d\theta}{2\pi}. \tag{4.29}$$

Let $n_f(r, H)$ be the number (with multiplicity counted) of zeros of $\langle \mathbf{f}, \mathbf{a} \rangle$ in $|z| < r$. Let $n_f^{(n)}(r, H)$ be the number of zeros of $\langle \mathbf{f}, \mathbf{a} \rangle$ in $|z| < r$, where the multiplicity is counted only as n if the vanishing order of $\langle \mathbf{f}, \mathbf{a} \rangle$ at the point is greater than or equal to n. The *counting function* $N_f(r, H)$ of f with respect to H is defined by

$$N_f(r,H) = \int_0^r (n_f(t,H) - n_f(0,H)) \frac{dt}{t} + n_f(0,H) \log r, \tag{4.30}$$

and the *truncated counting function* is

$$N_f^{(n)}(r,H) = \int_0^r (n_f^{(n)}(t,H) - n_f^{(n)}(0,H)) \frac{dt}{t} + n_f^{(n)}(0,H) \log r. \tag{4.31}$$

Note that the characteristic function, Weil function, proximity function, and the counting function introduced above are all independent of the choice of homogeneous coordinates.

By Corollary 4.1.3 (Jensen's formula)

$$N_f(r,H) = \int_0^{2\pi} \log |\langle \mathbf{f}(re^{i\theta}), \mathbf{a} \rangle| \frac{d\theta}{2\pi} + O(1).$$

So, according to the definitions, we derive the first main theorem:

$$T_f(r) = m_f(r,H) + N_f(r,H) + O(1).$$

To develop the theory, we first extend Proposition 4.1.6 to the higher-order derivatives.

Proposition 4.2.1. *Let* ϕ *be a non-constant meromorphic function in* $D(R_0)$ $(0 < R_0 \leq \infty)$. *Let* l *be a positive integer, and let* p, p' *be real numbers such that* $0 < pl < p' < 1$. *Let* $0 < r_0 < R_0$ *with* $T_\phi(r_0) \geq e$. *Then there exists a positive constant* K *such that for* $0 < r_0 < r < \rho < R_0$,

$$\int_0^{2\pi} \left| r^l \left(\frac{\phi^{(l)}}{\phi} \right) (re^{i\theta}) \right|^p \frac{d\theta}{2\pi} \leq K \left(\frac{\rho}{\rho - r} T_\phi(\rho) \right)^{p'}. \tag{4.32}$$

Proof. We use induction on l. Proposition 4.1.6 implies that our result holds for $l = 1$. Assume that the proposition holds for $l = k$, and we check it for $l = k+1$. Noticing that $\phi^{(k+1)} = (\phi^{(k)})'$, letting $p_1 = 1/(k+1)$ and $p_2 = k/(k+1)$, and applying Hölder's inequality, we get

$$\begin{aligned}
&\int_0^{2\pi}\left|r^{k+1}\left(\frac{\phi^{(k+1)}}{\phi}\right)(re^{i\theta})\right|^p\frac{d\theta}{2\pi}\\
&\leq \left(\int_0^{2\pi}\left|r^k\left(\frac{\phi^{(k)}}{\phi}\right)(re^{i\theta})\right|^{p/p_1}\frac{d\theta}{2\pi}\right)^{p_1}\left(\int_0^{2\pi}\left|r\left(\frac{(\phi^{(k)})'}{\phi^{(k)}}\right)(re^{i\theta})\right|^{p/p_2}\frac{d\theta}{2\pi}\right)^{p_2}\\
&\leq K\left(\frac{\rho}{\rho-r}T_\phi(\rho)\right)^{p'p_1}\left(\frac{\rho}{\rho-r}T_{\phi^{(k)}}(\rho)\right)^{p'p_2}. \qquad (4.33)
\end{aligned}$$

From the proof of Theorem 4.1.9 and from (4.19) we get

$$m_{\phi'/\phi}(r,\infty) \leq K\log^+\left(\frac{\rho}{\rho-r}T_\phi(r)\right).$$

Also,

$$N_{\phi'/\phi}(r,\infty) \leq N_\phi(r,\infty) + N_\phi(r,0) \leq 2T_\phi(r) + O(1).$$

Hence

$$\begin{aligned}
T_{\phi'}(r) &\leq T_{\phi'/\phi}(r) + T_\phi(r) + O(1) \leq m_{\phi'/\phi}(r,\infty) + N_{\phi'/\phi}(r,\infty) + T_\phi(r) + O(1)\\
&\leq 3T_\phi(r) + K\log^+\left(\frac{\rho}{\rho-r}T_\phi(r)\right) + O(1).
\end{aligned}$$

Using $\phi^{(k)} = (\phi^{(k)}/\phi^{(k-1)})\cdots(\phi'/\phi)$, we get

$$T_{\phi^{(k)}}(r) \leq C_k\left(T_\phi(r) + \log^+\left(\frac{\rho}{\rho-r}T_\phi(r)\right)\right) \qquad (4.34)$$

for some constant C_k, which may depend on k. For any $\epsilon > 0$, there exists a positive constant K_ϵ such that

$$\log^+\left(\frac{\rho}{\rho-r}T_\phi(r)\right) \leq K_\epsilon\left(\frac{\rho}{\rho-r}T_\phi(r)\right)^\epsilon.$$

Thus, for slightly smaller p' if necessary, by combining (4.33) and (4.34) we can conclude

$$\int_0^{2\pi}\left|r^{k+1}\left(\frac{\phi^{(k+1)}}{\phi}\right)(re^{i\theta})\right|^p\frac{d\theta}{2\pi} \leq K\left(\frac{\rho}{\rho-r}T_\phi(\rho)\right)^{p'}.$$

Thus the proposition is proved by induction. □

Let $f : D(R_0) \to \mathbb{P}^n(\mathbb{C})$ be a holomorphic map, where $0 < R_0 \le \infty$. Consider the hyperplanes

$$H_j : a_{j0}w_0 + \cdots + a_{jn}w_n = 0, \quad 1 \le j \le q.$$

We say that $H_1, \ldots, H_q$ are *in general position* if for any injective map $\mu : \{0, 1, \ldots, n\} \to \{1, \ldots, q\}$, $\mathbf{a}_{\mu(0)}, \ldots, \mathbf{a}_{\mu(n)}$ are linearly independent, where $\mathbf{a}_j = (a_{j0}, \ldots, a_{jn})$. Take hyperplanes

$$H_j : a_{j0}w_0 + \cdots + a_{jn}w_n = 0, \quad 1 \le j \le q,$$

which are in general position and such that $f(D(R_0)) \not\subset H_j$. We define the holomorphic functions

$$F_j = a_{j0}f_0 + \cdots + a_{jn}f_n, \quad 1 \le j \le q.$$

Denote by $W(f_0, \ldots, f_n)$ the Wronskian of $f_0, \cdots, f_n$. Assume that f is *linearly nondegenerate,* that is, its image is not contained in any proper subspace of $\mathbb{P}^n(\mathbb{C})$. Then $W(f_0, \ldots, f_n) \not\equiv 0$. We first prove the following theorem.

Theorem 4.2.2. *Let t, r_0 be positive numbers such that $0 < n(n+1)t/2 < p' < 1$ and $r_0 < R_0$. Assume that $T_f(r_0) \ge e$. Then there exists a positive constant K such that for $r_0 < r < \rho < R_0$,*

$$\int_0^{2\pi} \left| r^{\frac{n(n+1)}{2}} \frac{W(f_0, \ldots, f_n)}{F_1 \ldots F_q} \right|^t \|f\|^{t(q-n-1)}(re^{i\theta})d\theta \le K\left(\frac{\rho}{\rho - r} T_f(\rho)\right)^{p'}. \tag{4.35}$$

Proof. In the following, K_i $(i = 1, 2, \ldots)$ are some suitable positive constants. We first claim that there is a positive constant K_2 such that

$$\left|\frac{W(f_0, \ldots, f_n)}{F_1 F_2 \cdots F_q}\right| \|f\|^{q-n-1} \le K_2 \left(\sum_{1 \le i_1 < \cdots < i_{n+1} \le q} \left| \frac{W(F_{i_1}, \ldots, F_{i_{n+1}})}{F_{i_1} \cdots F_{i_{n+1}}} \right| \right). \tag{4.36}$$

Indeed, take an arbitrary point $z \in D(R_0)$. Let $i_1, \ldots, i_q$ be a permutation of the indices $1, 2, \ldots, q$ such that

$$|F_{i_1}(z)| \le \cdots \le |F_{i_{n+1}}(z)| \le |F_{i_{n+2}}(z)| \le \cdots \le |F_{iq}(z)|.$$

Since $H_1, \ldots, H_q$ are in general position, $f_0, \ldots, f_n$ are represented as linear combinations of $F_{i_1}, \ldots, F_{i_{n+1}}$, so e can find a positive constant $C_{i_1 \cdots i_{n+1}}$ independent of each z such that

$$|f_i(z)| \le C_{i_1 \cdots i_{n+1}} \max_{1 \le k \le n+1} |F_{i_k}(z)| \le C_{i_1 \cdots i_{n+1}} |F_{i_l}(z)|$$

for $i = 0, \dots, n$ and $l = n+1, \dots, q$. Thus

$$\|f(z)\| = \max_{0 \le i \le n} |f_i(z)| \le C_{i_1 \cdots i_{n+1}} |F_{i_l}(z)|$$

for $l = n+1, \dots, q$, and hence

$$\|f(z)\|^{q-n-1} \le K_1 |F_{i_{n+2}}(z) \cdots F_{i_q}(z)|.$$

On the other hand, we know that $W(f_0, \dots, f_n) = CW(F_{i_1}, \dots, F_{i_{n+1}})$ for a constant $C > 0$. Hence we conclude

$$\begin{aligned}
&\left| \frac{W(f_0, \dots, f_n)}{F_1 F_2 \cdots F_q}(z) \right| \|f(z)\|^{q-n-1} \\
&\le K_2 \left| \frac{W(F_{i_1}, \dots, F_{i_{n+1}}) F_{i_{n+2}} \cdots F_{i_q}}{F_1 F_2 \cdots F_q}(z) \right| \\
&= K_2 \left| \frac{W(F_{i_1}, \dots, F_{i_{n+1}})}{F_{i_1} \cdots F_{i_{n+1}}}(z) \right| \\
&\le K_2 \left(\sum_{1 \le i_1 < \cdots < i_{n+1} \le q} \left| \frac{W(F_{i_1}, \dots, F_{i_{n+1}})}{F_{i_1} \cdots F_{i_{n+1}}}(z) \right| \right).
\end{aligned}$$

This proves (4.36). Thus we only need to estimate

$$I_1 := \int_0^{2\pi} \left| \frac{W(F_{i_1}, \dots, F_{i_{n+1}})}{F_{i_1} \cdots F_{i_{n+1}}} \right|^t (re^{i\theta}) d\theta$$

by using the inequality $(\sum a_j)^t \le \sum a_j^t$ for $0 < t < 1$ and $a_j \ge 0$ for each j. Let

$$\chi := \frac{W(F_{i_1}, \dots, F_{i_{n+1}})}{F_{i_1} \cdots F_{i_{n+1}}}$$

and set $\varphi_j := F_{i_j} / F_{i_{n+1}}, 1 \le j \le n$. It is easy to see that

$$\chi = (-1)^n \det\left(\frac{\phi_i^{(l)}}{\phi_i} : 1 \le i, l \le n \right) = \sum_{(l_1, \dots, l_n)} (-1)^n \operatorname{sgn} \begin{pmatrix} 12 \cdots n \\ l_1 l_2 \cdots l_n \end{pmatrix} \frac{\varphi_1^{(l_1)}}{\varphi_1} \cdots \frac{\varphi_n^{(l_n)}}{\varphi_n},$$

where $(l_1, \dots, l_n)$ is a permutation of $(1, 2, \dots, n)$, and hence $l_1 + \cdots + l_n = \frac{n(n+1)}{2}$. Therefore the left-hand side of (4.35) can be estimated by a positive constant multiple of the sum of some functions of the form

$$\psi_{l_1 \dots l_n} := \left| r^{\frac{n(n+1)}{2}} \frac{\varphi_1^{(l_1)}}{\varphi_1} \cdots \frac{\varphi_n^{(l_n)}}{\varphi_n} \right|^t.$$

Put $s_j = l_j/(l_1 + \cdots + l_n)$. Then by the Hölder inequality we obtain

$$\frac{1}{2\pi}\int_0^{2\pi} \psi_{l_1\dots l_n}(re^{i\theta})d\theta \le \prod_{j=1}^{n}\left(\int_0^{2\pi}\left|r^{l_j}\left(\frac{\varphi_j^{(l_j)}}{\varphi_j}\right)(re^{i\theta})\right|^{t/s_j}\frac{d\theta}{2\pi}\right)^{s_j}.$$

Since $(t/s_j)l_j = t(l_1 + \cdots + l_n) = t\frac{n(n+1)}{2} < p' < 1$, we can apply Proposition 4.2.1 to get

$$\left(\int_0^{2\pi}\left|r^{l_j}\left(\frac{\varphi_j^{(l_j)}}{\varphi_j}\right)(re^{i\theta})\right|^{t/s_j}d\theta\right)^{s_j} \le K_3\left(\frac{\rho}{\rho - r}T_f(\rho)\right)^{p's_j}.$$

Hence

$$\begin{aligned}\frac{1}{2\pi}\int_0^{2\pi} \psi_{l_1\dots l_n}(re^{i\theta})d\theta &\le K_4\prod_{j=1}^{n}\left(\frac{\rho}{\rho - r}T_f(\rho)\right)^{p's_j}\\ &= K_4\left(\frac{\rho}{\rho - r}T_f(\rho)\right)^{p'}\end{aligned}$$

since $s_1 + \cdots + s_n = 1$. This completes the proof. □

Theorem 4.2.3 (Cartan's second main theorem). *Let f be a linearly nondegenerate holomorphic map of $D(R_0) = \{z \mid |z| < R_0\}(R_0 \le +\infty)$ into $\mathbb{P}^n(\mathbb{C})$, and let $H_1, \dots, H_q$ be hyperplanes in general position. Then, for $\epsilon > 0$, we have the inequality*

$$(q - (n+1))T_f(r) + N_W(r, 0) \le \sum_{j=1}^{q} N_f(r, H_j) + S(r) + \frac{n(n+1)}{2}\epsilon\log^+ r$$

for $0 < r < R_0$, where $S(r)$ is evaluated as follows:

(1) *In the case $R_0 < \infty$,*

$$S(r) \le C\left(\log^+\frac{1}{R_0 - r} + \log^+ T_f(r)\right)$$

for every $r \in [0, R_0)$ excluding a set E such that $\int_E \frac{1}{R_0 - t}dt < \infty$.

(2) *In the case $R_0 = \infty$, for every $\epsilon > 0$,*

$$S(r) \le \frac{n(n+1)}{2}((1+\epsilon)\log^+ T_f(r) + (1+\epsilon)^2\log^+\log T_f(r)) + O(1)$$

for every $r \in [0, +\infty)$ excluding a set E' such that $\int_{E'} dt < \infty$.

Proof. Let t, r_0 be positive numbers such that $0 < n(n+1)t/2 < p' < 1$ and $T_f(r_0) \ge e$ (note, if no such r_0 exists, then the theorem holds automatically). Then, by Theorem 4.2.2, for $r_0 < r < \rho < R_0$,

$$\int_0^{2\pi} \left|\frac{W(f_0,\ldots,f_n)}{F_1\ldots F_q}\right|^t \|f\|^{t(q-n-1)}(re^{i\theta})\frac{d\theta}{2\pi}$$
$$\leq K\left(\frac{\rho}{r(\rho-r)}T_f(\rho)\right)^{p'} \cdot r^{(p'-\frac{n(n+1)t}{2})}.$$

Hence by the concavity of the logarithm and the definition of $T_f(r)$ the above inequality implies that

$$t(q-n-1)T_f(r) + t\int_0^{2\pi} \log\left|\frac{W(f_0,\ldots,f_n)}{F_1\ldots F_q}\right|\frac{d\theta}{2\pi}$$
$$\leq p'\left(\log^+\frac{\rho}{r(\rho-r)} + \log^+ T_f(\rho) + O(1)\right) + \left(p' - \frac{n(n+1)t}{2}\right)\log r$$

for $r_0 < r < R < R_0$. Choose p' such that $(p'/t) < (1+\epsilon)\frac{n(n+1)}{2}$. Then

$$(q-n-1)T_f(r) + \int_0^{2\pi} \log\left|\frac{W(f_0,\ldots,f_n)}{F_1\ldots F_q}\right|\frac{d\theta}{2\pi}$$
$$\leq (1+\epsilon)\frac{n(n+1)}{2}\left(\log^+\frac{\rho}{r(\rho-r)} + \log^+ T_f(\rho) + O(1)\right) + \frac{n(n+1)}{2}\epsilon\log^+ r.$$

Denote

$$S(r) := (1+\epsilon)\frac{n(n+1)}{2}\left(\log^+\frac{\rho}{r(\rho-r)} + \log^+ T_f(\rho) + O(1)\right).$$

Then by Corollary 4.1.3 (Jensen's formula),

$$\int_0^{2\pi} \log\left|\frac{W(f_0,\ldots,f_n)}{F_1\ldots F_q}\right|\frac{d\theta}{2\pi} = N_W(r,0) - \sum_{j=1}^q N_f(r,H_j),$$

we get

$$(q-n-1)T_f(r) + N_W(r,0) \leq \sum_{j=1}^q N_f(r,H_j) + S(r) + \frac{n(n+1)}{2}\epsilon\log^+ r.$$

We now use Lemma 4.1.7 to estimate $S(r)$. Assume that $R_0 = \infty$. Take $\rho = r + \frac{1}{\log^{1+\epsilon} T_f(r)}$. Applying Lemma 4.1.7 with $F(r) = T_f(r)$, we have, for $r \geq r_0$ not contained in E such that $\int_E dt < \infty$, $\log T_f(\rho) \leq \log T_f(r) + 1$ and

$$\log^+\frac{\rho}{r(\rho-r)} \leq (1+\epsilon)\log^+\log T_f(r) + \log 2.$$

Hence

$$S(r) \leq_{\text{exc}} \frac{n(n+1)}{2}((1+\epsilon)\log^+ T_f(r) + (1+\epsilon)^2 \log^+ \log T_f(r)) + O(1),$$

where $\leq_{\text{exc}}$ means the inequality for all $r \notin E$. Similarly, in the case $R_0 < \infty$,

$$S(r) \leq C\left(\log^+ \frac{1}{R_0 - r} + \log^+ T_f(r)\right) + O(1)$$

for every $r \in [0, R_0)$ excluding a set E such that $\int_E \frac{1}{R_0 - t} dt < \infty$. □

Lemma 4.2.4. *Let $H_1, \dots, H_q$ be hyperplanes in $\mathbb{P}^n(\mathbb{C})$ located in general position. Then*

$$\sum_{j=1}^{q} N_f(r, H_j) - N_W(r, 0) \leq \sum_{j=1}^{q} N_f^{(n)}(r, H_j).$$

Proof. For each $z \in \mathbb{C}$, without loss of generality, we assume that $\langle \mathbf{f}, \mathbf{a}_j \rangle$ vanish at z for $1 \leq j \leq q_1$ and $\langle \mathbf{f}, \mathbf{a}_j \rangle$ do not vanish at z for $j > q_1$. There are integers $k_j \geq 0$ and nowhere vanishing holomorphic functions g_j in a neighborhood U of z such that

$$\langle \mathbf{f}, \mathbf{a}_j \rangle = (\zeta - z)^{k_j} g_j \quad \text{for } j = 1, \dots, q.$$

Here $k_j = 0$ if $q_1 < j \leq q$. Also, we can assume that $k_j \geq n$ if $1 \leq j \leq q_0$ and $1 \leq k_j < n$, where $0 \leq q_0 \leq q_1$. By the property of the Wronskian,

$$W = W(f_0, \dots, f_n) = CW(\langle \mathbf{f}, \mathbf{a}_{\mu(1)} \rangle, \dots \langle \mathbf{f}, \mathbf{a}_{\mu(n+1)} \rangle)$$

and

$$W(\langle \mathbf{f}, \mathbf{a}_{\mu(1)} \rangle, \dots, \langle \mathbf{f}, \mathbf{a}_{\mu(n+1)} \rangle) = \prod_{j=1}^{q_0} (\zeta - z)^{k_j - n} h(\zeta),$$

where $h(\zeta)$ is a holomorphic function on U. Thus W vanishes at z with order at least $\sum_{j=1}^{q_0}(k_j - n) = \sum_{j=1}^{q_0} k_j - q_0 n$. This, together with the definitions of $N_f(r, H_j)$, $N_W(r, 0)$, and $N_f^{(n)}(r, H_j)$, implies the lemma. □

Thus Theorem 4.2.3 can be reformulated as follows.

Theorem 4.2.5. *Let $f : D(R_0) \to \mathbb{P}^n(\mathbb{C})$, $0 < R_0 \leq \infty$, be a linearly nondegenerate holomorphic map, and let $H_1, \dots, H_q$ be hyperplanes in general position. Then, for any $\epsilon > 0$, we have the inequality*

$$(q - (n+1))T_f(r) \leq \sum_{j=1}^{q} N_f^{(n)}(r, H_j) + S(r) + \frac{n(n+1)}{2} \epsilon \log^+ r$$

for $0 < r < R_0$, *where* $S(r)$ *is evaluated as follows:*

(1) *In the case* $R_0 < \infty$,

$$S(r) \leq C\left(\log^+ \frac{1}{R_0 - r} + \log^+ T_f(r)\right)$$

for every $r \in [0, R_0)$ *excluding a set* E *such that* $\int_E \frac{1}{R_0 - t} dt < \infty$.

(2) *In the case* $R_0 = \infty$, *for every* $\epsilon > 0$,

$$S(r) \leq \frac{n(n+1)}{2}((1+\epsilon)\log^+ T_f(r) + (1+\epsilon)^2 \log^+ \log T_f(r)) + O(1)$$

for every $r \in [0, +\infty)$ *excluding a set* E' *such that* $\int_{E'} dt < \infty$.

Note that Theorem 4.2.2 assumes that $H_1, \ldots, H_q$ are in general position. Now we consider the case that $H_1, \ldots, H_q$ are not necessarily in general position. Let $K \subset \{1, \ldots, q\}$ be a set such that $\mathbf{a}_k, k \in K$, are linearly independent. Assuming that $\#K = n+1$, it is easy to prove the following: there is a constant C_1 such that

$$\left|\frac{W(f_0, \ldots, f_n)}{\prod_{j \in K}\langle \mathbf{f}, \mathbf{a}_j\rangle}\right| \|f\|^{q-n-1} \leq C_1\left(\sum_{1 \leq i_1 < \cdots < i_{n+1} \leq q}\left|\frac{W(F_{i_1}, \ldots, F_{i_{n+1}})}{F_{i_1} \cdots F_{i_{n+1}}}\right|\right),$$

where $F_j := \langle \mathbf{f}, \mathbf{a}_j\rangle$. With this, we can modify Theorem 4.2.2 with the same proof to get the following: *Let* $H_1, \ldots, H_q$ *be arbitrary hyperplanes in* $\mathbb{P}^n(\mathbb{C})$. *Let* $f = [f_0 : \cdots : f_n] : D(R_0) \to \mathbb{P}^n(\mathbb{C})$ *be a linearly nondegenerate holomorphic map with* $0 < R_0 \leq \infty$. *Let* t, r_0 *be positive numbers such that* $0 < n(n+1)t/2 < p' < 1$ *and* $r_0 < R_0$. *Then there exists a positive constant* C *such that for* $r_0 < r < \rho < R_0$,

$$\int_0^{2\pi} \max_K\left(\left|r^{\frac{n(n+1)}{2}} \frac{W(f_0, \ldots, f_n)}{\prod_{j \in K}\langle \mathbf{f}, \mathbf{a}_j\rangle}\right|^t \|f\|^{t(q-n-1)}(re^{i\theta})\right) d\theta \leq C\left(\frac{\rho}{\rho - r} T_f(\rho)\right)^{p'}, \tag{4.37}$$

where the maximum is taken over all subsets K *of* $\{1, \ldots, q\}$ *such that* $\#K = n+1$ *and* $\mathbf{a}_j$, $j \in K$, *are linearly independent.*

Theorem 4.2.6 (A general second main theorem [71]). *Let* $f = [f_0 : \cdots : f_n] : D(R_0) \to \mathbb{P}^n(\mathbb{C})$ *be a linearly nondegenerate holomorphic map with* $0 < R_0 \leq \infty$. *Let* $H_1, \ldots, H_q$ *(or* $\mathbf{a}_1, \ldots, \mathbf{a}_q$*) be arbitrary hyperplanes in* $\mathbb{P}^n(\mathbb{C})$. *Then for every* $\epsilon > 0$, *we have the inequality*

$$\int_0^{2\pi} \max_K \sum_{k \in K} \lambda_{H_k}(f(re^{i\theta}))\frac{d\theta}{2\pi} + N_W(r, 0) \leq (n+1)T_f(r) + S(r) + \frac{n(n+1)}{2}\epsilon \log^+ r$$

for $0 < r < R_0$, *where the maximum is taken over all subsets* K *of* $\{1, \ldots, q\}$ *such that* $\mathbf{a}_j, j \in K$, *are linearly independent, and* $S(r)$ *is the same as in Theorem 4.2.3.*

Proof. The proof is the same as that of Theorem 4.2.3, except using (4.37) instead of using Theorem 4.2.2. Also, note, without loss of generality, that we may assume that $q \geq n+1$ and $\#K = n+1$. □

Next, we extend Theorem 4.2.3 to the hypersurface case. A hypersurface $D \subset \mathbb{P}^n(\mathbb{C})$ is a zero set of a homogeneous polynomial of $n+1$ variables, that is,

$$D = \{[z_0 : \cdots : z_n] \in \mathbb{P}^n(\mathbb{C}) \mid Q(z_0, \ldots, z_n) = 0\},$$

where Q is a homogeneous polynomial of $n+1$ variables. The proximity function $m_f(r, D)$ and the counting function $N_f(r, D)$ can be defined similar to the hyperplane case as above. Let X be a projective variety. Divisors $D_1, \ldots, D_q$ on X are said to be *in general position* if for any subset $I \subseteq \{1, \ldots, q\}$ such that $\#I \leq n+1$,

$$\dim \bigcap_{i \in I} \operatorname{Supp}(D_i) \leq n - \#I,$$

where $\dim \emptyset = -1$.

Theorem 4.2.7 ([72]). *Let $f : \mathbb{C} \to \mathbb{P}^n(\mathbb{C})$ be an algebraically nondegenerate holomorphic map (i. e., its image $f(\mathbb{C})$ is not contained in any proper subvariety of $\mathbb{P}^n(\mathbb{C})$). Let $D_1, \ldots, D_q$ be hypersurfaces in $\mathbb{P}^n(\mathbb{C})$ of degree d_j located in general position. Then, for every $\epsilon > 0$ and for $l = 2^{n^2+4n} e^n d^{2n} (nI(\epsilon^{-1}))^n$, where d is the least common multiple of the d_j, we have*

$$(q - (n+1) - \epsilon) T_f(r) \leq \sum_{j=1}^{q} d_j^{-1} N_f^{(l-1)}(r, D_j),$$

where the inequality holds for all $r \in (0, +\infty)$ except for a possible set E of finite Lebesgue measure. Here $I(x) := \min\{a \in \mathbb{Z} \mid a > x\}$ for a real number x.

Define the defect of f for a hypersurface D of degree d:

$$\delta_{l-1}^{f,\star}(D) = 1 - \limsup_{r \to +\infty} \frac{N_f^{(l-1)}(r, D)}{d T_f(r)}.$$

Then Theorem 4.2.7 gives the following defect relation.

Corollary 4.2.8 (Defect relation). *Let $f : \mathbb{C} \to \mathbb{P}^n(\mathbb{C})$ be an algebraically nondegenerate holomorphic map, and let $D_1, \ldots, D_q$ be hypersurfaces in $\mathbb{P}^n(\mathbb{C})$ in general position. Then we have*

$$\sum_{j=1}^{q} \delta_{l-1}^{f,\star}(D_j) \leq n+1.$$

Before we prove Theorem 4.2.7, we first prove some preparation lemmas.

Lemma 4.2.9. *Let $\phi_1,\dots,\phi_m$ be homogeneous polynomials in $\mathbb{C}[X_0,\dots,X_n]$. Assume that they define a subvariety of $\mathbb{P}^n(\mathbb{C})$ of dimension $n-m$. Then $\{\phi_1,\dots,\phi_m\}$ is a regular sequence, that is, for $i=1,\dots,m$, the element ϕ_i is not a divisor of the zero in the ring $\frac{\mathbb{C}[X_0,\dots,X_n]}{(\phi_1,\dots,\phi_{i-1})}$.*

Proof. This is a well-known fact from the theory of Cohen–Macaulay rings. □

We will use the *lexicographic* ordering on the m-tuples $(i_1,\dots,i_m)\in\mathbb{N}^m$ of natural numbers. Namely, $(j_1,\dots,j_m)>(i_1,\dots,i_m)$ iff for some $b\in\{1,\dots,m\}$, we have $j_l=i_l$ for $l<b$ and $j_b>i_b$.

Lemma 4.2.10. *Let R be a commutative ring, and let $\phi_1,\dots,\phi_m$ be a regular sequence in R. Denote by $I\subset R$ the ideal generated by $\phi_1,\dots,\phi_m$. Suppose that for some $q,q_1,\dots,q_h\in R$, we have the equation*

$$\phi_1^{i_1}\cdots\phi_m^{i_m}q=\sum_{r=1}^{h}\phi_1^{j_1(r)}\cdots\phi_m^{j_m(r)}q_r,$$

where $(j_1(r),\dots,j_m(r))>(i_1,\dots,i_m)$ (in the lexicographic order) for $r=1,\dots,h$. Then $q\in I$.

Proof. We argue by induction on m. Since ϕ_1 is not a zero divisor in R, the statement is trivial for $m=1$. Assume that $m>1$ and the lemma is true up to $m-1$. Renumbering the indices $1,\dots,h$, we may assume that $j_1(r)>i_1$ for $r=1,\dots,s$ (possibly, $s=0$ or $s=h$) and that $j_1(r)=i_1$ for $r=s+1,\dots,h$ (the case $j_1(r)<i_1$ is excluded since $(j_1(r),\dots,j_m(r))>(i_1,\dots,i_m)$). Since ϕ_1 is not a zero divisor in R, we may then write

$$\phi_2^{i_2}\cdots\phi_m^{i_m}q=\phi_1\sigma+\sum_{r=s+1}^{h}\phi_2^{j_2(r)}\cdots\phi_m^{j_m(r)}q_r,$$

where $\sigma\in R$. We now reduce modulo ϕ_1, denoting the reduction with a bar, and work in the ring $R'=R/(\phi_1)$. We obtain

$$\bar\phi_2^{i_2}\cdots\bar\phi_m^{i_m}\bar q=\sum_{r=s+1}^{h}\bar\phi_2^{j_2(r)}\cdots\bar\phi_m^{j_m(r)}\bar q_r.$$

Note that $(j_2(r),\dots,j_m(r))>(i_2,\dots,i_m)$ for $r=s+1,\dots,m$ and that $\{\bar\phi_2,\dots,\bar\phi_m\}$ is a regular sequence in R'. We may thus apply the inductive assumption with $m-1$ in place of m and R' in place of R. We obtain that $\bar q$ lies in the ideal of R' generated by $\bar\phi_2,\dots,\bar\phi_m$, that is, $q\in I$, as required. This finishes the proof. □

Lemma 4.2.11. *Let V_N be the space of homogeneous polynomials of $n+1$ variables of degree N. Let $Q_1,\dots,Q_q$, $q\geq n+1$, be a set of homogeneous polynomials of $n+1$ variables of the same degree $d\geq 1$ and located in general position. Then for any nonnegative integer N and any $\{j_1,\dots,j_n\}\subset\{1,\dots,q\}$, the dimension of the vector space $\frac{V_N}{(Q_{j_1},\dots,Q_{j_n})\cap V_N}$ is equal to* *the number of n-tuples* $(\mathbf{i}) = (i_1, \dots, i_n) \in \mathbb{Z}^n_{\geq 0}$ *such that* $i_1 + \dots + i_n \leq N$ *and* $0 \leq i_1, \dots, i_n \leq d-1$*. In particular, for all* $N \geq n(d-1)$*, we have*

$$\dim \frac{V_N}{(Q_{j_1}, \dots, Q_{j_n}) \cap V_N} = d^n.$$

Proof. The case $N = 0$ holds trivially, so we assume that N is positive for the rest of the proof. We first prove that

$$\dim \frac{V_N}{(Q_{j_1}, \dots, Q_{j_n}) \cap V_N} = \dim \frac{V_N}{(Q_1, \dots, Q_n) \cap V_N} \tag{4.38}$$

for any choice of $\{j_1, \dots, j_n\} \subset \{1, \dots, q\}$ and any N. For this, it suffices to prove that

$$\dim((Q_1, \dots, Q_n) \cap V_N) = \dim((Q_{j_1}, \dots, Q_{j_n}) \cap V_N).$$

Since the order of the Q_j does not matter, we only need to prove that

$$\dim((Q_1, \dots, Q_n) \cap V_N) = \dim((Q_1, \dots, Q_{n-1}, Q_{j_n}) \cap V_N), \tag{4.39}$$

and the rest follows by induction. Notice that to prove (4.39), we only need to show that

$$\dim \frac{(Q_1, \dots, Q_n) \cap V_N}{(Q_1, \dots, Q_{n-1}) \cap V_N} = \dim \frac{(Q_1, \dots, Q_{n-1}, Q_{j_n}) \cap V_N}{(Q_1, \dots, Q_{n-1}) \cap V_N}. \tag{4.40}$$

To do so, consider the map

$$\phi : \frac{(Q_1, \dots, Q_n) \cap V_N}{(Q_1, \dots, Q_{n-1}) \cap V_N} \to \frac{(Q_1, \dots, Q_{n-1}, Q_{j_n}) \cap V_N}{(Q_1, \dots, Q_{n-1}) \cap V_N}$$
$$\left[\sum_{j=1}^{n-1} b_j Q_j + b_n Q_n\right] \mapsto \left[\sum_{j=1}^{n-1} b_j Q_j + b_n Q_{j_n}\right]$$

with $b_j \in \mathbb{C}[x_0, \dots, x_n]$. This map is clearly surjective, so we only need to prove that it is well defined and injective. To prove that ϕ is well defined, let $[\sum_{j=1}^{n-1} b_j Q_j + b_n Q_n] = [\sum_{j=1}^{n-1} b_j Q_j + b'_n Q_n]$. This means that $(b_n - b'_n)Q_n \in (Q_1, \dots, Q_{n-1}) \cap V_N$. Since $Q_1, \dots, Q_n$ is a regular sequence by Lemma 4.2.9, Q_n is not a zero divisor in $\frac{\mathbb{C}[x_0, \dots, x_n]}{(Q_1, \dots, Q_{n-1})}$. Hence

$$\left[\sum_{j=1}^{n-1} b_j Q_j + b_n Q_{j_n}\right] - \left[\sum_{j=1}^{n-1} b_j Q_j + b'_n Q_{j_n}\right] = [(b_n - b'_n)Q_{j_n}] = 0$$

in $\frac{(Q_1, \dots, Q_{n-1}, Q_{j_n}) \cap V_N}{(Q_1, \dots, Q_{n-1}) \cap V_N}$. So ϕ is well defined. The injectivity of ϕ follows by the same argument, just changing the roles of Q_n and Q_{j_n}, and using the fact that $Q_1, \dots, Q_{n-1}, Q_{j_n}$ is also a regular sequence by Lemma 4.2.9. Hence we proved that (4.40) is true, and thus (4.38) holds. We remark that for the proof, we only used the fact that $Q_1, \dots, Q_n, Q_{j_n}$ are in general position.

Since $Q_1, \dots, Q_n$ define a subvariety of dimension 0, there exists a hyperplane H_{n+1} on $\mathbb{P}^n(\mathbb{C})$ such that $\bigcap_{j=1}^n \{Q_j = 0\} \cap H_{n+1} = \emptyset$. Furthermore, by induction there exist hyperplanes $H_1, \dots, H_{n+1}$ such that $\bigcap_{j=1}^{i-1} \{Q_j = 0\} \cap (\bigcap_{k=i}^n H_k) = \emptyset$. This means that $Q_1, \dots, Q_{i-1}, H_i^d, \dots, H_n^d$ are in general position. Then by (4.39) and by induction we get that

$$\dim \frac{V_N}{(Q_1, \dots, Q_n) \cap V_N} = \dim \frac{V_N}{(H_1^d, \dots, H_n^d) \cap V_N}.$$

Let $L_1, \dots, L_{n+1}$ be the linear forms defining $H_1, \dots, H_{n+1}$. Since $L_1, \dots, L_{n+1}$ are linearly independent, it follows from a well-known fact of linear algebra that there exists a permutation $\{k_1, \dots, k_{n+1}\}$ of $\{1, \dots, n+1\}$ such that $\{L_1, \dots, L_{i-1}, x_{k_i}, \dots, x_{k_{n+1}}\}$ are linearly independent for any $i \in \{1, \dots, n+2\}$. This means that $\{L_1^d, \dots, L_{i-1}^d, x_{k_i}^d, \dots, x_{k_{n+1}}^d\}$ are in general position. Hence by (4.39) and by induction we get that

$$\dim \frac{V_N}{(L_1^d, \dots, L_n^d) \cap V_N} = \dim \frac{V_N}{(x_1^d, \dots, x_n^d) \cap V_N},$$

and thus

$$\dim \frac{V_N}{(Q_{j_1}, \dots, Q_{j_n}) \cap V_N} = \dim \frac{V_N}{(x_1^d, \dots, x_n^d) \cap V_N}.$$

On the other hand, it is easy to see that for any positive integer N, the vector space $\frac{V_N}{(x_1^d, \dots, x_n^d) \cap V_N}$ has a basis $\{x_0^{N-(i_1+\cdots+i_n)} x_1^{i_1} \cdots x_n^{i_n} \mid i_1 + \cdots + i_n \le N, 0 \le i_1, \dots, i_n \le d-1\}$. This completes the proof. □

Let $D_1, \dots, D_q$ be hypersurfaces in $\mathbb{P}^n(\mathbb{C})$ located in general position. Let $Q_j, 1 \le j \le q$, be the homogeneous polynomials in $\mathbb{C}[X_0, \dots, X_n]$ of degree d_j defining D_j. Replacing Q_j by Q_j^{d/d_j} if necessary, where d is the l. c. m. of the d_j, we can assume that $Q_1, \dots, Q_q$ have the same degree d. For $N \in \mathbb{N}$, let V_N be the space of homogeneous polynomials of degree N of $n+1$ variables. Pick n distinct polynomials $\gamma_1, \dots, \gamma_n \in \{Q_1, \dots, Q_q\}$. Arrange the n-tuples $\mathbf{i} = (i_1, \dots, i_n)$ of nonnegative integers by lexicographic order. For the n-tuples $\mathbf{i} = (i_1, \dots, i_n)$ of nonnegative integers with $\sigma(\mathbf{i}) := \sum_j i_j \le N/d$, define the spaces

$$W_{\mathbf{i}} = W_{N,\mathbf{i}} := \sum_{\mathbf{e} \ge \mathbf{i}} \gamma_1^{e_1} \cdots \gamma_n^{e_n} V_{N-d\sigma(\mathbf{e})}. \tag{4.41}$$

Clearly, $W_{(0,\dots,0)} = V_N$ and $W_{\mathbf{i}} \supset W_{\mathbf{i}'}$ if $\mathbf{i}' \ge \mathbf{i}$, so that $\{W_{\mathbf{i}}\}$ in fact defines a filtration of V_N. Suppose that $\mathbf{i}'$ follows $\mathbf{i}$ in the ordering. We have the following lemma.

Lemma 4.2.12. *There is an isomorphism*

$$\frac{W_{\mathbf{i}}}{W_{\mathbf{i}'}} \cong \frac{V_{N-d\sigma(\mathbf{i})}}{(\gamma_1,\dots,\gamma_n)\cap V_{N-d\sigma(\mathbf{i})}}.$$

Proof. Define the vector space homomorphism $\phi : V_{N-d\sigma(\mathbf{i})} \to \frac{W_{\mathbf{i}}}{W_{\mathbf{i}'}}$ as follows: for a polynomial $q(X_0,\dots,X_n)$ in $V_{N-d\sigma(\mathbf{i})}$, we define $\phi(q)$ as the class of $\gamma_1^{i_1}\cdots\gamma_n^{i_n}q$ (which belongs to $W_{\mathbf{i}}$) modulo $W_{\mathbf{i}'}$. By our definition of the spaces $W_{\mathbf{e}}$ this homomorphism is surjective. To find its kernel, suppose that $q \in \ker\phi$. This means that $\gamma_1^{i_1}\cdots\gamma_n^{i_n}q$ lies in $\sum_{(\mathbf{e})>(\mathbf{i})}\gamma_1^{e_1}\cdots\gamma_n^{e_n}V_{N-d\sigma(\mathbf{e})}$. Then we may write

$$\gamma_1^{i_1}\cdots\gamma_n^{i_n}q = \sum_{(\mathbf{e})>(\mathbf{i})}\gamma_1^{e_1}\cdots\gamma_n^{e_n}q_{(\mathbf{e})}$$

for elements $q_{(\mathbf{e})} \in V_{N-d\sigma(\mathbf{e})}$. By Lemma 4.2.10, q lies in the ideal generated by $\gamma_1,\dots\gamma_n$. Therefore $q = \sum_{j=1}^n a_j\gamma_j$, where $a_j, 1 \le j \le n$, are homogeneous with $\deg a_j = \deg q - d$. Then $a_j \in V_{N-d(\sigma(\mathbf{i})+1)}$. Thus we see that $\gamma_1^{i_1}\cdots\gamma_n^{i_n}q$ is a sum of terms in $W_{\mathbf{i}'}$, which concludes the proof of the lemma. □

Let $\triangle_{\mathbf{i}} = \dim(W_{\mathbf{i}}/W_{\mathbf{i}'})$, where $\mathbf{i}' > \mathbf{i}$ are consecutive n-tuples with $W_{\mathbf{i}'} \subset W_{\mathbf{i}}$. By Lemma 4.2.11, $\triangle_{\mathbf{i}} = d^n$ for every $\mathbf{i}$ such that $N - d\sigma(\mathbf{i}) \ge n(d-1)$. Moreover, Lemma 4.2.11 implies that $\triangle_{\mathbf{i}}$ is independent of the choice of $\gamma_1,\dots,\gamma_n$. Hence $\sum_{\mathbf{i}} i_j\triangle_{\mathbf{i}}$ is independent of the choice of $\gamma_1,\dots,\gamma_n$ and of $j = 1,\dots,n$. Ffor $1 \le j \le n$, set

$$\triangle := \sum_{\mathbf{i}} i_j\,\triangle_{\mathbf{i}}\,. \tag{4.42}$$

Lemma 4.2.13. *With* $N = 2d(n+1)(nd+n)(2^n-1)(I(\epsilon^{-1})+1)+nd$ *for any* $\epsilon > 0$, *we have*

$$\frac{lN}{\triangle} \le d(n+1) + \epsilon/2,$$

where $l = \binom{N+n}{n}$. *Moreover, l satisfies the estimate*

$$l \le 2^{n^2+4n}e^n d^{2n}(nI(\epsilon^{-1}))^n,$$

where $I(x) := \min\{k \in \mathbb{N} : k > x\}$ *for a positive real number* x.

Proof. First, notice that

$$l = \binom{N+n}{n} = \frac{(n+N)(n+N-1)\cdots(N+1)N!}{N!n!} \le \frac{(N+n)^n}{n!}.$$

Now since N is divisible by d, it follows from Lemma 4.2.11 that

$$\triangle = \sum_{\sigma(\mathbf{i})\le N/d} i_j\triangle_{\mathbf{i}} \ge \sum_{\sigma(\mathbf{i})\le N/d-n} i_j\triangle_{\mathbf{i}} = d^n \sum_{\sigma(\mathbf{i})\le N/d-n} i_j$$
$$= \frac{d^n}{n+1} \sum_{\sigma(\hat{\mathbf{i}})=N/d-n} \sum_{j=1}^{n+1} i_j = \frac{d^n}{n+1} \sum_{\sigma(\hat{\mathbf{i}})=N/d-n} (N/d-n)$$
$$= \frac{d^n}{n+1}\binom{N/d}{n}(N/d-n) = \frac{N(N-d)\cdots(N-dn)}{d(n+1)!},$$

where $\hat{\mathbf{i}} = (i_1, \dots, i_{n+1})$, and we used the fact that the number of nonnegative integer m-tuples with sum $\le T$ for a positive integer T is equal to the number of nonnegative integer $(m+1)$-tuples with sum exactly T, that is, to $\binom{T+m}{m}$.

Since $(N-dj) \ge (N-dn)$ for every integer $j \le n$, we have

$$\prod_{j=1}^{n} \frac{1}{N-dj} \le \left(\frac{1}{N-dn}\right)^n,$$

and thus

$$\frac{lN}{\triangle} \le d(n+1)\left(\frac{N+n}{N-nd}\right)^n.$$

Using

$$N = 2d(n+1)(nd+n)(2^n-1)(I(\epsilon^{-1})+1)+nd,$$

we find that

$$\left(\frac{N+n}{N-nd}\right)^n = \left(1+\frac{n+nd}{N-nd}\right)^n = 1+\sum_{r=1}^{n}\binom{n}{r}\left(\frac{nd+n}{N-nd}\right)^r$$
$$\le 1+(2^n-1)\frac{nd+n}{N-nd} \le 1+\frac{\epsilon}{2d(n+1)}.$$

Therefore

$$\frac{lN}{\triangle} \le d(n+1)+\epsilon/2.$$

To estimate l, we use the inequality

$$\binom{x+y}{y} \le \frac{(x+y)^{x+y}}{x^x y^y} = \left(1+\frac{y}{x}\right)^x\left(1+\frac{x}{y}\right)^y \le \left(e\left(1+\frac{x}{y}\right)\right)^y$$

for positive integers x, y. Hence, with $N = 2d(n+1)(nd+n)(2^n-1)(I(\epsilon^{-1})+1)+nd$, we have

$$l = \binom{N+n}{n} \le e^n\left(1+\frac{N}{n}\right)^n$$
$$\le e^n(1+2d(n+1)(d+1)(2^n-1)(I(\epsilon^{-1})+1)+d)^n$$
$$\le 2^{n^2+4n}e^n d^{2n}(nI(\epsilon^{-1}))^n.$$

This proves the lemma. □

Proof of Theorem 4.2.7. Let $Q_j, 1 \le j \le q$, be the homogeneous polynomials in $\mathbb{C}[X_0, \dots, X_n]$ of degree d_j defining D_j. Replacing Q_j by Q_j^{d/d_j} if necessary, where d is the l. c. m. of d_j, we can assume that $Q_1, \dots, Q_q$ have the same degree of d. Let $N \in \mathbb{N}$, and let V_N be the space of homogeneous polynomials of $n+1$ variables of degree N and fix an (arbitrary) basis $\phi_1, \dots, \phi_l$, where $l = \dim V_N$. Set $F = [\phi_1(f) : \cdots : \phi_l(f)]$. Since f is algebraically nondegenerate, $F : \mathbb{C} \to \mathbb{P}^{l-1}(\mathbb{C})$ is linearly nondegenerate. Given $z \in \mathbb{C}$, there exists a numbering $\{i_1, \dots, i_q\}$ of the indices $1, \dots, q$ such that

$$|(Q_{i_1} \circ f)(z)| \le \cdots \le |(Q_{i_q} \circ f)(z)|.$$

Since $Q_1, \dots, Q_q$ are in general position, Hilbert Nullstellensatz implies that for any integer $k, 0 \le k \le n$, there is an integer $m_k \ge d$ such that

$$x_k^{m_k} = \sum_{j=1}^{n+1} b_{jk}(x_0, \dots, x_n) Q_{i_j}(x_0, \dots, x_n),$$

where $b_{jk}, 1 \le j \le n+1, 0 \le k \le n$, are homogeneous forms with coefficients in $\mathbb{C}$ of degree $m_k - d$. So

$$|f_k(z)|^{m_k} \le c_1 \|f(z)\|^{m_k - d} \max\{|Q_{i_1}(f)(z)|, \dots, |Q_{i_{n+1}}(f)(z)|\},$$

where c_1 is a positive constant depending only on the coefficients of b_{jk} and thus depends only on the coefficients of Q_j. Therefore

$$\|f(z)\|^d \le c_1 \max\{|Q_{i_1}(f)(z)|, \dots, |Q_{i_{n+1}}(f)(z)|\}.$$

Therefore we derive that

$$\prod_{j=1}^{q} \frac{\|f(z)\|^d}{|Q_j(f)(z)|} \le c_1^{q-n} \prod_{k=1}^{n} \frac{\|f(z)\|^d}{|Q_{i_k}(f)(z)|}. \tag{4.43}$$

Take $\gamma_1 = Q_{i_1}, \dots, \gamma_n = Q_{i_n}$ and let $W_{N,\mathbf{i}}$ be defined by (4.41). Then we get a filtration $V_N = W_\mathbf{0} \supset \cdots \supset W_\mathbf{i} \supset W_{\mathbf{i}'} \supset \cdots$, where we just write $W_\mathbf{i}$ for $W_{N,\mathbf{i}}$. We now choose a basis $\psi_1, \dots, \psi_l$ for V_N in the following way: We start with the last nonzero $W_\mathbf{i}$ and pick a basis of it; Then we continue inductively as follows: suppose $\mathbf{i}' > \mathbf{i}$ are consecutive n-tuples such that $d\sigma(\mathbf{i}), d\sigma(\mathbf{i}') \le N$ and assume that we have chosen a basis of $W_{(\mathbf{i}')}$;

It follows directly from the definition that we may pick representatives in $W_{\mathbf{i}}$ for the quotient space $W_{\mathbf{i}}/W_{\mathbf{i}'}$, of the form $\gamma_1^{i_1}\cdots\gamma_n^{i_n}\eta$, where $\eta\in V_{N-d\sigma(\mathbf{i})}$. We extend the previously constructed basis in $W_{\mathbf{i}'}$ by adding these representatives. In particular, we have obtained a basis for $W_{\mathbf{i}}$, and our induction procedure may go on unless $W_{\mathbf{i}}=V_N$. Note that if we let ψ be an element of the basis constructed with respect to $W_{\mathbf{i}}/W_{\mathbf{i}'}$, then we may write $\psi=\gamma_1^{i_1}\cdots\gamma_n^{i_n}\eta$, where $\eta\in V_{N-d\sigma(\mathbf{i})}$. Thus we have the bound

$$|\psi(f)(z)|\le c_2|\gamma_1(f)(z)|^{i_1}\cdots|\gamma_n(f)(z)|^{i_n}\|f(z)\|^{N-d\sigma(i)},$$

where c_2 is a positive constant depending only on f and $Q_1,\dots,Q_q$. Observe that there are precisely $\triangle_{\mathbf{i}}$ such functions ψ in our basis. Write $\psi_1,\dots,\psi_l$ as linear forms $L_1,\dots,L_l$ in $\phi_1,\dots,\phi_l$ such that $\psi_t(f)=L_t(F)$, where $F=[\phi_1(f):\cdots:\phi_l(f)]$. Then from the above we have

$$\prod_{t=1}^{l}|L_t(F(z))|\le K\Big(\prod_{\mathbf{i}=(i_1,\dots,i_n)}|\gamma_1^{i_1}(f(z))\cdots\gamma_n^{i_n}(f(z))|^{\triangle_{\mathbf{i}}}\Big)\|f(z)\|^{lN-d\sum_{\mathbf{i}}\sigma(\mathbf{i})\triangle_{\mathbf{i}}},$$

where, as we noted earlier, K is a constant depending only on f and $D_1,\dots,D_q$ and may be different each time. So

$$\frac{\|f(z)\|^{d\sum_{\mathbf{i}}\sigma(\mathbf{i})\triangle_{\mathbf{i}}}}{\prod_{\mathbf{i}}|\gamma_1^{i_1\triangle_{\mathbf{i}}}(f(z))|\cdots|\gamma_n^{i_n\triangle_{\mathbf{i}}}(f(z))|}\le K\frac{\|f(z)\|^{lN}}{\prod_{t=1}^{l}|L_t(F(z))|}.$$

Thus using the definition that $\triangle:=\sum_{\mathbf{i}}i_j\triangle_{\mathbf{i}}$ for each $1\le j\le n$, we get

$$\frac{\|f(z)\|^{dn\triangle}}{|\gamma_1^{\triangle}(f(z))|\cdots|\gamma_n^{\triangle}(f(z))|}\le K\frac{\|f(z)\|^{lN}}{\prod_{t=1}^{l}|L_t(F(z))|}.$$

With $\gamma_1=Q_{i_1},\dots,\gamma_n=Q_{i_n}$, it gives

$$\frac{\|f(z)\|^{dn\triangle}}{|Q_{i_1}^{\triangle}(f(z))\cdots Q_{i_n}^{\triangle}(f(z))|}\le K\frac{\|f(z)\|^{lN}}{\prod_{t=1}^{l}|L_t(F(z))|}.$$

On the other hand, from (4.43) we have

$$\frac{\|f(z)\|^{dq\triangle}}{|Q_1^{\triangle}(f(z))\cdots Q_q^{\triangle}(f(z))|}\le K\frac{\|f(z)\|^{dn\triangle}}{|Q_{i_k}^{\triangle}(f)(z)\cdots Q_{i_n}^{\triangle}(f)(z)|}.$$

Hence, also using the fact that $c_1\|f(z)\|^N\le\|F(z)\|\le c_2\|f(z)\|^N$, we derived the following key estimate:

$$\frac{\|f(z)\|^{dq\triangle}}{|Q_1^{\triangle}(f(z))\cdots Q_q^{\triangle}(f(z))|}\le C\frac{\|F(z)\|^{l}}{\prod_{t=1}^{l}|L_t(F(z))|},\tag{4.44}$$

where C is a constant.

Note that although $L_1, \dots, L_l$ depend on z, there are only finitely many such choices, and we denote the total collection as $\{L_1, \dots, L_u\}$, since there are only finitely many choices of $\{\gamma_1, \dots \gamma_n\} \subset \{Q_1, \dots, Q_q\}$. Estimate (4.44) implies that

$$\triangle \sum_{j=1}^{q} m_f(r, D_j) \leq \int_0^{2\pi} \max_K \log \prod_{j\in K} \frac{\|F(re^{i\theta})\| \|L_j\|}{|L_j(F)(re^{i\theta})|} \frac{d\theta}{2\pi} + O(1),$$

where $\max_K$ is taken over all subsets K of $\{1, \dots, u\}$ such that the linear forms $L_j, j \in K$, are linearly independent. Applying Theorem 4.2.6 (with $R_0 = \infty$) with $\epsilon/(2N)$ to the holomorphic map F and linear forms $L_1, \dots, L_u$ and using the fact that $T_F(r) \leq NT_f(r) + O(1)$, we obtain

$$\begin{aligned}\int_0^{2\pi} \max_K \log \prod_{j\in K} \frac{\|F(re^{i\theta})\| \|L_j\|}{|L_j(F)(re^{i\theta})|} \frac{d\theta}{2\pi} &\leq_{\text{exc}} (l + \epsilon/(2N))T_F(r) - N_W(r, 0)\\ &\leq (lN + \epsilon/2)T_f(r) - N_W(r, 0),\end{aligned}$$

where $\leq_{\text{exc}}$ means the inequality for all r outside of a set E of finite Lebesgue measure, and where $W = W(F_0, \dots, F_l)$. Hence

$$\triangle \sum_{j=1}^{q} m_f(r, D_j) \leq_{\text{exc}} (lN + \epsilon)T_f(r) - N_W(r, 0).$$

Using the first main theorem this gives

$$\left(qd - \frac{lN}{\triangle}\right) T_f(r) \leq_{\text{exc}} \sum_{j=1}^{q} N_f(r, Q_j) - \frac{1}{\triangle} N_W(r, 0) + (\epsilon/2)T_f(r).$$

From Lemma 4.2.13 with $N = 2d(n+1)(nd+n)(2^n - 1)(I(\epsilon^{-1}) + 1) + nd$ for any $\epsilon > 0$, we have

$$\frac{lN}{\triangle} \leq d(n+1) + \epsilon/2,$$

where $l = \binom{N+n}{n}$. Moreover, l satisfies the estimate

$$l \leq 2^{n^2+4n} e^n d^{2n} (nI(\epsilon^{-1}))^n,$$

where $I(x) := \min\{k \in \mathbb{N} : k > x\}$ for a positive real number x. With such l, we get

$$(d(q - (n+1))T_f(r) \leq_{\text{exc}} \sum_{j=1}^{q} N_f(r, Q_j) - \frac{1}{\triangle} N_W(r, 0) + \epsilon T_f(r).$$

We now estimate $\sum_{j=1}^{q} N_f(r,Q_j) - \frac{1}{\triangle}N_W(r,0)$. For each $z \in \mathbb{C}$, without loss of generality, we assume that $Q_j(f)$ vanishes at z for $1 \le j \le q_1$ and $Q_j(f)$ does not vanish at z for $j > q_1$. By the hypothesis that the D_j are in general position we know that $q_1 \le n$. There are integers $k_j \ge 0$ and nowhere vanishing holomorphic functions g_j such that for $j = 1, \ldots, q$,

$$Q_j(f) = (\zeta - z)^{k_j} g_j,$$

where $k_j = 0$ if $q_1 < j \le q$. For $\{Q_1, \ldots, Q_n\} \subset \{Q_1, \ldots, Q_q\}$, as above, we can obtain a basis $\psi_1, \ldots, \psi_l$ of V_N and linearly independent linear forms $L_1, \ldots, L_l$ such that $\psi_t(f) = L_t(F), 1 \le t \le l$. From basic properties of Wronskians we have

$$W = W(F_1, \ldots, F_l) = CW(L_1(F), \ldots, L_l(F)) = CW(\psi_1(f), \ldots, \psi_l(f)),$$

where C is some constant. Let ψ be an element of the basis $\{\psi_1, \ldots, \psi_l\}$. As we discussed earlier, we may write $\psi = Q_1^{i_1} \cdots Q_n^{i_n} \eta$ with $\eta \in V_{N-d\sigma(\mathbf{i})}$. Therefore

$$\psi(f) = (Q_1(f))^{i_1} \cdots (Q_n(f))^{i_n} \eta(f).$$

Note that there are $\triangle_{\mathbf{i}}$ such that ψ is our basis. Notice that $(Q_j(f))^{i_j} = (\zeta - z)^{i_j k_j} g_j^{i_j}$ for $j = 1, \ldots, n$. Also, we can assume that $k_j \ge l-1$ for $1 \le j \le q_0$ and $1 \le k_j < l-1$ for $q_0 < j \le q_1$. Then W vanishes at z with order at least

$$\begin{aligned}
&\sum_{\mathbf{i}} \left(\sum_{j=1}^{q_0} i_j (k_j - (l-1)) \right) \triangle_{\mathbf{i}} \\
&\quad = \sum_{j=1}^{q_0} \left[\left(\sum_{\mathbf{i}} i_j \triangle_{\mathbf{i}} \right) (k_j - (l-1)) \right] = \triangle \sum_{j=1}^{q_0} (k_j - (l-1)).
\end{aligned}$$

On the other hand,

$$\sum_{j=1}^{q} k_j = \sum_{j=1}^{n} k_j = \sum_{j=1}^{q_0} k_j + \sum_{j=q_0}^{q_1} k_j.$$

Hence

$$\sum_{j=1}^{q} N_f(r,Q_j) - \frac{1}{\triangle} N_W(r,0) \le \sum_{j=0}^{q} N_f^{[l-1]}(r,D_j).$$

The theorem is thus proved. □

Finally, we extend the above result by replacing $\mathbb{P}^n(\mathbb{C})$ with an arbitrary complex projective variety X.

Theorem 4.2.14 ([73]). *Let X be a complex projective variety of dimension $n \geq 1$. Let $D_1, \ldots, D_q$ be hypersurfaces on X located in general position. Assume that each D_j, $1 \leq j \leq q$, is linearly equivalent to d_jA, where A is an ample divisor. Let $f : \mathbb{C} \to X$ be an algebraically nondegenerate holomorphic map. Then, for every $\epsilon > 0$,*

$$\sum_{j=1}^{q} d_j^{-1} m_f(r, D_j) \leq_{\text{exc}} (n+1+\epsilon) T_f(r),$$

where $\leq_{\text{exc}}$ means the inequality for all $r \in (0, +\infty)$ except for a possible set E of finite Lebesgue measure.

To prove Theorem 4.2.14, we first introduce some notation. Let $Y \subset \mathbb{P}^M(\mathbb{C})$ be a projective variety of dimension n (we will take $M = q-1$ in the proof). Denote by $\mathbb{C}[x_0, \ldots, x_M]_m$ the vector space of homogeneous polynomials in $\mathbb{C}[x_0, \ldots, x_M]$ of degree m (including $m = 0$). Let I_Y be the prime ideal in $\mathbb{C}[x_0, \ldots, x_M]$ defining Y. Put $(I_Y)_m = \mathbb{C}[x_0, \ldots, x_M]_m \cap I_Y$. The *Hilbert polynomial* $H_Y(m)$ of Y is defined by

$$H_Y(m) := \dim(\mathbb{C}[x_0, \ldots, x_M]_m / (I_Y)_m).$$

The *Hilbert weight* of Y with respect to the weight $\mathbf{c} \in \mathbb{R}^M$ is defined by

$$S_Y(m, \mathbf{c}) = \max\left(\sum_{i=1}^{H_Y(m)} \mathbf{a}_i \cdot \mathbf{c} \right), \tag{4.45}$$

where the maximum is taken over all sets of monomials $\mathbf{x}^{\mathbf{a}_1}, \ldots, \mathbf{x}^{\mathbf{a}_{H_Y(m)}}$ whose residue classes modulo I_Y form a basis of $\mathbb{C}[x_0, \ldots, x_N]_m / (I_Y)_m$, where for $\mathbf{x} = (x_0, \ldots, x_M)$ and $\mathbf{a} = (a_0, \ldots, a_M) \in \mathbb{Z}_+^{M+1}$, $\mathbf{x}^{\mathbf{a}} = x_0^{a_0} \cdots x_M^{a_M}$. We need the following lemma to estimate (get a lower bound) of Hilbert weight $S_Y(m, \mathbf{c})$.

Lemma 4.2.15 ([73]). *Let $Y \subset \mathbb{P}^M$ be an algebraic variety of dimension n and degree $\triangle$. Let $m > \triangle$ be an integer, and let $c = (c_0, \ldots, c_M) \in \mathbb{R}^{M+1}$. Let $\{i_0, \ldots, i_n\}$ be a subset of $\{0, \ldots, M\}$ such that*

$$\{x = [x_0 : \cdots : x_M] \in \mathbb{P}^M \mid x_{i_0} = \cdots = x_{i_n} = 0\} \cap Y = \emptyset.$$

Then

$$\frac{1}{mH_Y(m)} S_Y(m, \mathbf{c}) \geq \frac{1}{(n+1)} (c_{i_0} + \cdots + c_{i_n}) - \frac{(2n+1)\triangle}{m} \left(\max_{0 \leq i \leq M} c_i \right).$$

Proof of Theorem 4.2.14. We assume that A is very ample, and we embed X through A into the complex projective space $X \subset \mathbb{P}^N(\mathbb{C})$. So we can assume that $D_1, \ldots, D_q$ are defined by $Q_1, \ldots, Q_q$, where each Q_j is a homogeneous polynomial of degree d_j in $N+1$ variables. Replacing Q_j by Q_j^{d/d_j} if necessary, where d is the l. c. m. of the d_j, we can assume that $Q_1, \ldots, Q_q$ have the same degree of d. Consider the map $\chi : X \to \mathbb{P}^{q-1}(\mathbb{C})$ given by $\chi(w) = [Q_1(w) : \cdots : Q_q(w)]$. Then it is a finite morphism since $D_1, \ldots D_q$ are in general position. Let $Y = \chi(X)$. Then Y is a complex subvariety of $\mathbb{P}^{q-1}(\mathbb{C})$ with $\dim Y = \dim X = n$ and

$$\triangle := \deg Y \le d^n \deg X.$$

Let m be a positive integer. Consider the vector space $V_m = \mathbb{C}[x_1, \ldots, x_q]_m/(I_Y)_m$, and fix a basis $\phi_0, \ldots, \phi_{n_m}$, where $n_m = H_Y(m) - 1$. Consider the map

$$F = [\phi_0(\chi \circ f), \ldots, \phi_{n_m}(\chi \circ f)] : \mathbb{C} \to \mathbb{P}^{n_m}.$$

For each $z \in \mathbb{C}$, define

$$c_j(z) = \log \frac{\|f(z)\|^d \|Q_j\|}{|Q_j(f)(z)|}, \quad 1 \le j \le q,$$

and let $\mathbf{c}(z) = (c_1(z), \ldots, c_q(z))$. With this given weight $\mathbf{c}(z)$, take monomials (which depend on z) $\mathbf{x}^{\mathbf{a}_1}, \ldots, \mathbf{x}^{\mathbf{a}_{H_Y(m)}}$ whose residue classes modulo I_Y form a basis of $V_m = \mathbb{C}[x_1, \ldots, x_q]_m/(I_Y)_m$ and which reaches the maximum in the Hilbert weight $S_Y(m, \mathbf{c}(z))$ (see (4.45)). For each $1 \le j \le H_Y(m)$, write $\mathbf{x}^{\mathbf{a}_j} = L_{j,z}(\phi_0, \ldots, \phi_{n_m})$. Then $L_{j,z}$ are linear forms that are linearly independent for each z. Note that the linear forms depend on z, but there are only finitely many of them in total. We prove the following key estimate: For each $z \in \mathbb{C}$ and for the linear forms $L_{j,z}, 1 \le j \le H_Y(m)$, chosen above, we have

$$\begin{aligned}\sum_{j=1}^{q} \log \frac{\|f(z)\|^d \|Q_j\|}{|Q_j(f)(z)|} \le{}& \frac{(n+1)}{mH_Y(m)} \sum_{j=1}^{H_Y(m)} \log \frac{\|F(z)\|}{|L_{j,z}(F)(z)|} + d(n+1) \log\|f(z)\| \\ &- \frac{n+1}{m} \log\|F(z)\| + \frac{(n+1)(2n+1)\triangle}{m} \left(\max_{0 \le i \le q} \log \frac{\|f(z)\|^d \|Q_i\|}{|Q_i(f)(z)|} \right) \\ &+ O(1). \end{aligned} \tag{4.46}$$

Indeed, since $D_1, \ldots, D_q$ are in general position, for every $z \in \mathbb{C}$, there are $i_0, \ldots, i_n$ (which may depend on z) such that

$$\sum_{j=1}^{q} \log \frac{\|f(z)\|^d \|Q_j\|}{|Q_j(f)(z)|} \le \sum_{t=0}^{n} \log \frac{\|f(z)\|^d \|Q_{i_t}\|}{|Q_{i_t}(f)(z)|}.$$

With $c_j(z) = \log \frac{\|f(z)\|^d \|Q_j\|}{|Q_j(f)(z)|}$ and $\mathbf{c}(z) = (c_1(z), \ldots, c_q(z))$, by applying Lemma 4.2.15 we get

$$\frac{1}{mH_Y(m)} S_Y(m, \mathbf{c}(z)) \ge \frac{1}{(n+1)} (c_{i_0}(z) + \cdots + c_{i_n}(z)) - \frac{(2n+1)\triangle}{m} \left(\max_{1 \le i \le q} c_i(z) \right),$$

that is,

$$\sum_{t=0}^{n}\log\frac{\|f\|^d\|Q_{i_t}\|}{|Q_{i_t}(f)(z)|}\le\frac{(n+1)S_Y(m,\mathbf{c}(z))}{mH_Y(m)}+\frac{(n+1)(2n+1)\triangle}{m}\left(\max_{0\le i\le q}\log\frac{\|f(z)\|^d\|Q_i\|}{|Q_i(f)(z)|}\right).$$

Therefore

$$\sum_{j=1}^{q}\log\frac{\|f(z)\|^d\|Q_j\|}{|Q_j(f)(z)|}\le\frac{(n+1)S_Y(m,\mathbf{c}(z))}{mH_Y(m)}+\frac{(n+1)(2n+1)\triangle}{m}\left(\max_{0\le i\le q}\log\frac{\|f(z)\|^d\|Q_i\|}{|Q_i(f)(z)|}\right).$$

However, by our selection of the linear forms $L_{j,z}$, $L_{j,z}(\phi_0,\dots,\phi_{n_m})=\mathbf{x}^{\mathbf{a}_j}$, and

$$S_Y(m,\mathbf{c}(z))=\sum_{i=1}^{H_Y(m)}(\mathbf{a}_i\cdot\mathbf{c}(z)).$$

Hence

$$L_{i,z}(F(z))=(Q_1(f)(z))^{a_{i1}}\cdots(Q_q(f)(z))^{a_{iq}},$$

where $\mathbf{a}_i=(a_{i1},\dots,a_{iq})$. Thus

$$\begin{aligned}-\log|L_{i,z}(F(z))|&=-(a_{i1}\log|Q_1(f)(z)|+\cdots+a_{iq}\log|Q_q(f)(z)|)\\&=a_{i1}c_1(z)+\cdots+a_{iq}c_q(z)-m\log\|f(z)\|^d+O(1)\\&=\mathbf{a}_i\cdot\mathbf{c}(z)-m\log\|f(z)\|^d+O(1).\end{aligned}$$

Therefore

$$-\sum_{i=1}^{H_Y(m)}\log|L_{i,z}(F(z))|=S_Y(m,\mathbf{c}(z))-mH_Y(m)\log\|f(z)\|^d+O(1).$$

Hence

$$\begin{aligned}\sum_{j=1}^{q}\log\frac{\|f(z)\|^d\|Q_j\|}{|Q_j(f)(z)|}&\le\frac{(n+1)}{mH_Y(m)}\sum_{i=1}^{H_Y(m)}\log\frac{\|F(z)\|}{|L_{i,z}(F(z))|}+d(n+1)\log\|f(z)\|\\&\quad-\frac{n+1}{m}\log\|F(z)\|+\frac{(n+1)(2n+1)\triangle}{m}\left(\max_{0\le i\le q}\log\frac{\|f(z)\|^d\|Q_i\|}{|Q_i(f)(z)|}\right)+O(1).\end{aligned}$$

This proves (4.46).

We now continue our proof. Integrating (4.46), applying Theorem 4.2.6 (with $R_0=\infty$) with $\epsilon=1$, and using the first main theorem and the fact that $T_F(r)\le mdT_f(r)+O(1)$, we have

$$\sum_{j=1}^{q}m_f(r,D_j)$$

$$
\begin{aligned}
&\leq_{\text{exc}} \frac{(n+1)(H_Y(m)+1)}{mH_Y(m)} T_F(r) + d(n+1)T_f(r) \\
&\quad - \frac{n+1}{m} T_F(r) + \frac{(2n+1)(n+1)\triangle}{m} \int_{1\leq j\leq q} \left(\max_{1\leq i\leq q} \log \frac{\|f(re^{i\theta})\|^d \|Q_i\|}{|Q_i(f)(re^{i\theta})|} \right) \frac{d\theta}{2\pi} + O(1) \\
&\leq \frac{d(n+1)}{H_Y(m)} T_f(r) + d(n+1)T_f(r) \\
&\quad + \frac{(2n+1)(n+1)\triangle}{m} \int_{1\leq j\leq q} \left(\max_{1\leq i\leq q} \log \frac{\|f(re^{i\theta})\|^d \|Q_i\|}{|Q_i(f)(re^{i\theta})|} \right) \frac{d\theta}{2\pi} + O(1).
\end{aligned}
$$

For each $j \in \{1, \dots, q\}$, by Jensen's formula,

$$
\begin{aligned}
\int_0^{2\pi} \log \frac{\|f(re^{i\theta})\|^d}{|Q_j(f)(re^{i\theta})|} \frac{d\theta}{2\pi} &= d \int_0^{2\pi} \log\|f(re^{i\theta})\| \frac{d\theta}{2\pi} - \int_0^{2\pi} \log|Q_j(f)(re^{i\theta})| \frac{d\theta}{2\pi} \\
&\leq dT_f(r) - N_f(r, D_j) + O(1).
\end{aligned}
$$

Choose m big enough such that

$$
\frac{(2n+1)(n+1)q\triangle}{m} < \frac{\epsilon}{4d}, \quad \text{and} \quad \frac{(n+1)d}{H_Y(m)} < \frac{\epsilon}{4}.
$$

Then we have

$$
\sum_{j=1}^{q} m_f(r, D_j) \leq_{\text{exc}} d(n+1+\epsilon)T_f(r).
$$

This proves the theorem. □

4.3 The degenerate case

In Theorem 4.2.3, we assumed that the map is linearly nondegenerate. In this subsection, we deal with the degenerated case, that is, the image of $f : D(R_0) \to \mathbb{P}^n(\mathbb{C})$ is contained in a proper subspace of $\mathbb{P}^n(\mathbb{C})$, where $0 < R_0 \leq \infty$. We assume that the image $f : D(R_0) \to \mathbb{P}^n(\mathbb{C})$ is contained in a subspace of dimension k but not in any subspace of dimension lower than k. Without loss of generality, we assume that the subspace of dimension k that contains $f(\mathbb{C})$ is $\mathbb{P}^k(\mathbb{C})$. Then $f : D(R_0) \to \mathbb{P}^k(\mathbb{C})$ is linearly nondegenerate. However, the difficulty in applying the results above is that hyperplanes $H_1, \dots, H_q$ in $\mathbb{P}^n(\mathbb{C})$ in general position may not necessarily be in general position after being restricted to $\mathbb{P}^k(\mathbb{C})$. Nochka [53, 54] (see also [11]) developed a technique to overcome this difficulty.

Let $n \geq k$ and $q \geq n+1$. We consider the hyperplanes $H_j, 1 \leq j \leq q$, in $\mathbb{P}^k(\mathbb{C})$ given by

$$H_j = \{[x_0 : \cdots : x_k] \in \mathbb{P}^k(\mathbb{C}) \mid a_{j0}x_0 + \cdots + a_{jk}x_k = 0\},$$

with reduced nonzero coefficient vectors $\mathbf{a}_j = (a_{j0}, \ldots, a_{jk}) \in \mathbb{C}^{k+1^*}$, where $\mathbb{C}^{k+1^*}$ is the dual space of $\mathbb{C}^{k+1}$.

Definition 4.3.1. The hyperplanes $H_1, \ldots, H_q$ (or $\mathbf{a}_1, \ldots, \mathbf{a}_q$) in $\mathbb{P}^k(\mathbb{C})$ are said to be in *n-subgeneral position* if for all $1 \leq i_0 < \cdots < i_n \leq q$, the linear span of $\mathbf{a}_{i_0}, \ldots, \mathbf{a}_{i_n}$ is $\mathbb{C}^{k+1^*}$.

Given a collection of nonzero vectors $A = \{\mathbf{a}_1, \ldots, \mathbf{a}_q\}$ in $\mathbb{C}^{k+1^*}$ in n-subgeneral position. We introduce the *Nochka diagram* and *Nochka polygon* of A: For a nonempty subset B of A, we associate with B a point $P_B = (\#B, d(B))$ in $\mathbb{R}^2$, where $d(B)$ is the dimension of the linear span of B. The collection of the points $\{P_B \mid B \subset A\}$ is called the *Nochka diagram* of A (see Figure 4.1). In Figure 4.1 the point $O = (0,0)$, $U = (n-k, 0)$, $V = (n+1, k+1)$, $W = (\frac{2n-k+1}{2}, \frac{k+1}{2})$, and $X = (2n-k+1, k+1)$, so W is the midpoint of the segment UV, as well as of the segment OX. By the n-subgeneral position assumption, the points $P_B = (\#B, d(B))$ with $\#B \geq n+1$ lie on the line with $d(B) = k+1$ (i. e., they lie on the horizontal line through V and to the right of V). On the other hand, the points $P_B = (\#B, d(B))$ with $\#B \leq n+1$ lie on or above the line through U and V.

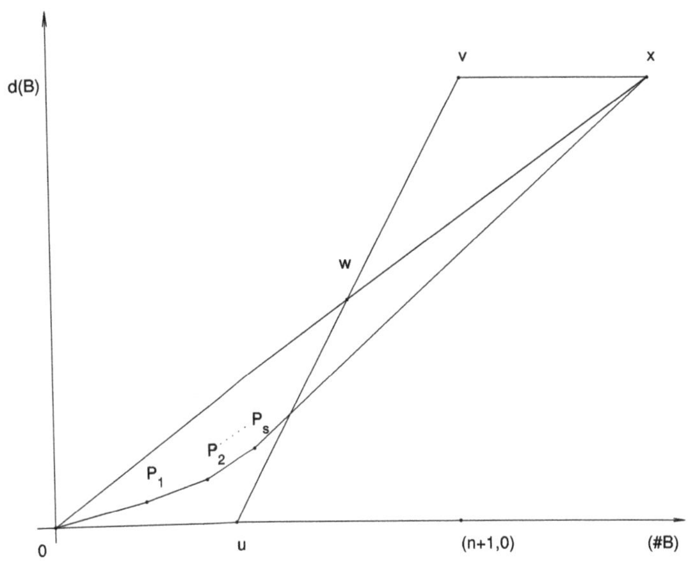

Figure 4.1: Nochka diagram.

Proposition 4.3.2. *Let $A = \{\mathbf{a}_1, \ldots, \mathbf{a}_q\}$ be a set of vectors in $\mathbb{C}^{k+1^*}$ in n-subgeneral position. Then either $\sigma(O, P_B) \geq \sigma(O, X)$ for all $P_B = (\#B, d(B))$ with $\#B \leq n+1$, where σ denotes the slope of the associated line segments, or there exists a uniquely determined sequence of subsets of A,*

$$A \supset B_s \supset B_{s-1} \supset \cdots \supset B_1 \supset B_0 = \emptyset,$$

with the following properties:
(i) $\sigma(P_{B_{i-1}}, P_{B_i}) < \sigma(P_{B_{i-1}}, X), 1 \leq i \leq s$, *where σ denotes the slope of the associated line segments, and $X = (2n - k + 1, k + 1)$.*
(ii) $\sigma(O, P_{B_i}) < \sigma(O, X)$ *for $1 \leq i \leq s$, where $O = (0, 0)$ is the origin, and X is as above.*
(iii) $\sigma(P_{B_{i-1}}, P_{B_i}) < \sigma(P_{B_i}, P_{B_{i+1}}), 1 \leq i \leq s$, *where we set $P_{B_{s+1}} = X$.*
(iv) *For $0 \leq i \leq s$, let $\mathcal{U}_i$ be the collection of sets $B \subset A$ strictly containing B_i with $\#B \leq n+1$. Then $\sigma(P_{B_i}, P_{B_{i+1}}) \leq \sigma(P_{B_i}, P_B)$ for any $B \in \mathcal{U}_i$ with strict inequality if $i < s$ and $B \in \mathcal{U}_{i+1}$.*

Proof. We construct the sets $B_1, \ldots, B_s$ inductively. Suppose that $B_0, \ldots, B_j$ have been constructed. Then by induction hypothesis, (i) and (ii) are satisfied for all $0 \leq i \leq j$ (these conditions are empty if $j = 0$), and (iii) and (iv) are satisfied for $0 \leq i < j$ (these conditions are empty if $j \leq 1$).

If $\sigma(P_{B_j}, P_B) \geq \sigma(P_{B_j}, X)$ for all $B \in \mathcal{U}_j$, that is, (iv) is satisfied for $i = j$, then we set $j = s$. By (i) of the induction hypothesis we have $\sigma(P_{B_{s-1}}, P_{B_s}) < \sigma(P_{B_{s-1}}, X)$. This implies that (consider the triangle $P_{B_{s-1}} P_{B_s} X$) $\sigma(P_{B_{s-1}}, P_{B_s}) < \sigma(P_{B_s}, X)$, which is (iii) for the case $i = j = s$.

We may now assume that there exists $B \in \mathcal{U}_j$ such that

$$\sigma(P_{B_j}, P_B) < \sigma(P_{B_j}, X). \tag{4.47}$$

Let

$$\sigma_j = \min_{B \in \mathcal{U}_j} \{\sigma(P_{B_j}, P_B)\}$$

and

$$\mathcal{M}_j = \{B \in \mathcal{U}_j \mid \sigma(P_{B_j}, P_B) = \sigma_j\}.$$

For $B \in \mathcal{M}_j$, we claim that $d(B) < (k+1)/2$. In fact, $\sigma(P_{B_j}, P_B) < \sigma(P_{B_j}, X)$ since $B \in \mathcal{M}_j$ By (ii) we also have $\sigma(O, P_{B_j}) < \sigma(O, X)$. These conditions and the remark before the proposition imply that P_B lies in the triangle OUW that appears in the Nochka diagram but not on the segment OW. Hence $d(B) < (k+1)/2$.

We now claim that if B and C are in $\mathcal{M}_j$, then $B \cup C$ is also in $\mathcal{M}_j$. First of all, since $d(B \cup C) \leq d(B) + d(C) < k+1$, the subgeneral position condition implies that $\#(B \cup C) < n+1$. Thus $(B \cup C) \in \mathcal{U}_j$. Now

$$\begin{aligned} d(B \cup C) - d(B_j) &= d(B) + d(C) - d(B \cap C) - d(B_j) \\ &= d(B) - d(B_j) + d(C) - d(B_j) - d(B \cap C) + d(B_j) \\ &\leq \sigma_j\{\#(B) - \#(B_j) + \#(C) - \#(B_j) - \#(B \cap C) + \#(B_j)\} \\ &= \sigma_j\{\#(B \cup C) - \#(B_j)\}, \end{aligned}$$

that is, $\sigma(P_{B_j}, P_{B \cup C}) \leq \sigma_j$. Thus $(B \cup C) \in \mathcal{M}_j$. The claim is proved.

We now define

$$B_{j+1} = \bigcup_{B \in \mathcal{M}_j} B.$$

The above claim implies that $B_{j+1} \in \mathcal{M}_j$. We now check (i)–(iv). The proceeding argument shows that $B_{j+1} \in \mathcal{M}_j$. By construction, $\sigma_j = \sigma(P_{B_j}, P_{B_{j+1}}) \leq \sigma(P_{B_j}, P_B)$ for all $B \in \mathcal{U}_j$. The inequality is strict if B is in $\mathcal{U}_{j+1}$. Thus (iv) is verified for $i = j$. By assumption (4.47), $\sigma(P_{B_j}, P_{B_{j+1}}) < \sigma(P_{B_j}, X)$, so (i) is verified for $i = j+1$. This, together with $\sigma(O, P_{B_j}) < \sigma(O, X)$ (by induction hypothesis, (ii) holds for $i \leq j$), implies that the point $P_{B_{j+1}}$ is below the line OX. Hence $\sigma(O, P_{B_{j+1}}) < \sigma(O, X)$, which is (ii) for $i = j + 1$. From (iv) with $i = j - 1$ and $B = B_{j+1}$ we have $\sigma(P_{B_{j-1}}, P_{B_j}) < \sigma(P_{B_{j-1}}, P_{B_{j+1}})$. This implies that (by considering the triangle $P_{B_{j-1}} P_{B_j} P_{B_{j+1}}$) $\sigma(P_{B_{j-1}}, P_{B_j}) < \sigma(P_{B_j}, P_{B_{j+1}})$. Thus (iii) is verified for $i = j$.

This completes the induction step. Since the sets $B_0, B_1, \ldots$ are strictly increasing and A is a finite set, the above construction terminates after a finite number of steps, concluding the proof of the proposition. □

By Proposition 4.3.2, either $\sigma(O, P_B) \geq \sigma(O, X)$ for all $P_B = (\#B, d(B))$ with $\#B \leq n + 1$, or there exists a sequence $B_1, \ldots, B_s$ that appears in the proposition. If the first case occurs, then we define $\omega(\mathbf{a}) = 1$ for $\mathbf{a} \in A$. Otherwise, the sequence $B_1, \ldots, B_s$ that appears in the proposition gives rise to a polygon in $\mathbb{R}^2$, which is called the *Nochka polygon* of A (see Fig. 4.1). Set $B_{s+1} = A$. Then $(B_1 - B_0) \cup (B_2 - B_1) \cup \cdots \cup (B_s - B_{s-1}) \cup (A - B_s) = A$ is a partition of A. We now define the *Nochka weights* $\omega(\mathbf{a})$ of $\mathbf{a} \in A$ as follows: For any $\mathbf{a} \in A$, $\mathbf{a}$ lies in $B_{i+1} - B_i$ for some $0 \leq i \leq s$, and we define

$$\omega(\mathbf{a}) = \sigma(P_{B_i}, P_{B_{i+1}}) \tag{4.48}$$

and

$$\theta = \frac{1}{\sigma(P_{B_s}, X)} = \frac{2n - k + 1 - \#B_s}{k + 1 - d(B_s)}. \tag{4.49}$$

The significance of Nochka weights is shown by the following theorem.

Theorem 4.3.3 (Nochka's theorem). *Let $H_1, \ldots, H_q$ (or $\mathbf{a}_1, \ldots, \mathbf{a}_q$) be hyperplanes in $\mathbb{P}^k(\mathbb{C})$ in n-subgeneral position with $q \geq 2n - k + 1$. Then there exists a function $\omega : \{1, \ldots, q\} \to \mathbb{R}(0, 1]$, called a Nochka weight, and a real number $\theta \geq 1$, called the Nochka constant, satisfying the following properties:*

(i) *If* $j \in \{1, \ldots, q\}$, *then* $0 \leq \omega(j)\theta \leq 1$.
(ii) $q - 2n + k - 1 = \theta(\sum_{j=1}^{q} \omega(j) - k - 1)$.
(iii) *If* $\emptyset \neq B \subset \{1, \ldots, q\}$ *with* $\#B \leq n + 1$, *then* $\sum_{j \in B} \omega(j) \leq \dim L(B)$, *where* $L(B)$ *is the linear space generated by* $\{\mathbf{a}_j \mid j \in B\}$.
(iv) $1 \leq (n+1)/(k+1) \leq \theta \leq (2n - k + 1)/(k + 1)$.
(v) *For any real numbers* $E_1, \ldots, E_q$ *such that* $E_j \geq 1$ *for* $1 \leq j \leq q$ *and any* $Y \subset \{1, \ldots, q\}$ *with* $0 < \#Y \leq n+1$, *there exists a subset* M *of* Y *with* $\#M = \dim L(Y)$ *such that* $\{\mathbf{a}_j\}_{j \in M}$ *is a basis for the linear space* $L(Y)$ *generated by* $\{\mathbf{a}_j \mid j \in Y\}$ *and*

$$\prod_{j \in Y} E_j^{\omega(j)} \leq \prod_{j \in M} E_j.$$

Proof. With the $\omega(j)$ and θ defined by (4.48) and (4.49), (i) immediately follows from (iii) of Proposition 4.3.2. To verify (ii), we set $B_{s+1} = \{1, 2, \ldots, q\}$. Then denoting $\sigma_i = \sigma(P_{B_i}, P_{B_{i+1}})$, we have

$$\begin{aligned}\sum_{j=1}^{q} \omega(j) &= \sum_{0 \leq i \leq s} \sigma_i(\#B_{i+1} - \#B_i) \\ &= \sum_{0 \leq i \leq s-1} \{d(B_{i+1}) - d(B_i)\} + \sigma_s(q - \#B_s) \\ &= d(B_s) + \sigma_s(q - \#B_s),\end{aligned}$$

where, by definition,

$$\sigma_s = \sigma(P_{B_s}, X) = \frac{k + 1 - d(B_s)}{2n - k + 1 - \#B_s} = \frac{1}{\theta}.$$

Hence

$$d(B_s) = k + 1 - \sigma_s(2n - k + 1 - \#B_s).$$

Substituting yields (ii).

For (iii), we consider the two cases (a) $\#(B \cup B_s) \geq n + 1$ and (b) $\#(B \cup B_s) < n + 1$. If $\#(B \cup B_s) \geq n + 1$, then the subgeneral position implies

$$k + 1 \leq d(B \cup B_s). \tag{4.50}$$

By property (i) the Nochka weights satisfy $\sigma_s^{-1}\omega(\mathbf{a}) \leq 1$. Thus

$$\sum_{\mathbf{a} \in B} \omega(\mathbf{a}) \leq \sigma_s \#B.$$

By the definition of subgeneral position any set $\#B \leq n + 1$, say $\#B = n + 1 - p$, satisfies $k + 1 - p \leq d(B)$. Thus

$$\#B \le d(B) + n - k.$$

Therefore

$$\sum_{\mathbf{a}\in B} \omega(\mathbf{a}) \le \sigma_s(d(B) + n - k) = d(B)\sigma_s\left(1 + \frac{n-k}{d(B)}\right).$$

By (4.50), $k + 1 \le d(B) + d(B_s)$, so that

$$\begin{aligned}\sum_{\mathbf{a}\in B} \omega(\mathbf{a}) &\le \sigma_s(d(B) + n - k) = d(B)\sigma_s\left(1 + \frac{n-k}{d(B)}\right)\\ &\le d(B)\sigma_s\left(1 + \frac{n-k}{k+1-d(B_s)}\right)\\ &= d(B)\sigma_s\frac{n+1-d(B_s)}{k+1-d(B_s)}\\ &\le d(B)\sigma_s\frac{2n-k+1-\#B_s}{k+1-d(B_s)}\\ &= d(B).\end{aligned}$$

So (iii) holds in case (a).

We now assume that $\#(B \cup B_s) < n + 1$. Then the set $B'_{i+1} = B_i \cup (B \cap B_{i+1})$ contains B_i, and $\#B'_{i+1} \le \#(B \cup B_s) < n + 1$ for all i. It follows that B'_{i+1} is in $\mathcal{U}_i$, and by part (iv) of Proposition 4.3.2

$$\sigma_i = \sigma(P_{B_i}, P_{B_{i+1}}) \le \sigma(P_{B_i}, P_{B'_{i+1}}) = \frac{d(B_i \cup (B \cap B_{i+1})) - d(B_i)}{\#(B_i \cup (B \cap B_{i+1})) - \#B_i}.$$

Since B_{i+1} contains B_i, we have $\#(B_i \cup (B \cap B_{i+1})) - \#B_i = \#(B \cap B_{i+1}) - \#(B \cap B_i)$ and $d(B_i \cup (B \cap B_{i+1})) - d(B_i) = d(B \cap B_{i+1}) - d(B \cap B_i)$. Thus

$$\sigma_i \le \frac{d(B \cap B_{i+1}) - d(B \cap B_i)}{\#(B \cap B_{i+1}) - \#(B \cap B_i)}.$$

We can now easily estimate the sum of Nochka weights:

$$\begin{aligned}\sum_{\mathbf{a}\in B} \omega(\mathbf{a}) &\le \sum_{0\le i\le s} \sigma_i\{\#(B \cap B_{i+1}) - \#(B \cap B_i)\}\\ &\le \sum_{0\le i\le s} (d(B \cap B_{i+1}) - d(B \cap B_i))\\ &= d(B \cap B_{s+1}) = d(B).\end{aligned}$$

This completes the proof of (iii).

To verify (iv), since P_{B_s} lies below the line OX in the Nochka diagram, $\sigma_s = \sigma(P_{B_s}, X) > \sigma(O, X) = (k + 1)/(2n - k + 1)$. On the other hand, P_{B_s} lies below the triangle OUW, and thus $\sigma(P_{B_s}, X) < \sigma(U, X) = (k + 1)/(n + 1)$. This proves (iv).

Finally, we prove (v). Without loss of generality, we assume that $1 \leq E_q \leq E_{q-1} \leq \cdots \leq E_1$. Define an increasing sequence of subsets of Y as follows: Let $i_1 = \min\{i \mid i \in Y\}$ and $I_1 = \{i \in Y \mid \mathbf{a}_i$ is a multiple of $\mathbf{a}_{i_1}\}$. If $Y - I_1 \neq \emptyset$, then choose $i_2 \in Y - I_1$ such that $i_2 = \min\{i \mid i \in Y - I_1\}$ and $I_2 = \{i \in Y \mid \mathbf{a}_i \in$ linear span of $\mathbf{a}_{i_1}, \mathbf{a}_{i_2}\}$. Inductively, if I_{j-1} is defined and if $Y - I_{j-1} \neq \emptyset$, then $i_j = \min\{i \mid i \in Y - I_{j-1}\}$ and $I_j = \{i \in Y \mid \mathbf{a}_i \in$ linear span of $\mathbf{a}_{i_1}, \ldots, \mathbf{a}_{i_j}\}$. This process stops at I_p with $p =$ the dimension of $L(Y)$. It is clear that

$$I_p \supset I_{p-1} \supset \cdots$$

and $i_p \geq i_{p-1} \geq \cdots \geq i_1$. Let $M = \{i_1, \ldots, i_p\}$. Then, by construction, the set $\{\mathbf{a}_j\}_{j \in M}$ is a basis for $L(Y)$. Set $I_0 = \emptyset$. Then $Y = \bigcup_{1 \leq j \leq p}(I_j - I_{j-1})$ is a disjoint union. Since $E_q \leq \cdots \leq E_1$, by construction we have

$$E_{i_j} = \max_{i \in I_j - I_{j-1}} E_i.$$

Thus

$$\prod_{j \in Y} E_j^{\omega(j)} \leq \prod_{1 \leq j \leq p} \prod_{i \in I_j - I_{j-1}} E_i^{\omega(i)} \leq \prod_{1 \leq j \leq p} E_{i_j}^{\alpha_j},$$

where $\alpha_j = \sum_{i \in I_j - I_{j-1}} \omega(i)$. Since the sets I_j are increasing, for any $1 \leq r \leq p$, (iii) implies that

$$\sum_{1 \leq j \leq r} \alpha_j = \sum_{1 \leq j \leq r} \sum_{i \in I_j - I_{j-1}} \omega(i) = \sum_{i \in I_r} \omega(i) \leq d(I_r) = r. \tag{4.51}$$

It remains to show that

$$\prod_{1 \leq j \leq p} E_{i_j}^{\alpha_j} \leq \prod_{1 \leq j \leq p} E_{i_j}.$$

This is easily verified by the induction on p. For $p = 1$, by (iii), $\alpha_1 \leq 1$, and since $E_1 \geq 1$, we trivially have $E_1^{\alpha_1} \leq E_1$. Assume that the inequality holds for $p = k$. Since $\alpha_{k+1} \leq k + 1 - \sum_{1 \leq j \leq k} \alpha_j$ by (4.51), we have

$$\prod_{1 \leq j \leq k+1} E_{i_j}^{\alpha_j} \leq \left(\prod_{1 \leq j \leq k} \left(\frac{E_{i_j}}{E_{i_{k+1}}} \right)^{\alpha_j} \right) E_{i_{k+1}}^{k+1}.$$

Since $i_p \geq \cdots \geq i_1$, we have $1 \leq E_{i_j}/E_{i_{k+1}}$ for $1 \leq j \leq k$, and the induction hypothesis implies that

$$\left(\prod_{1 \leq j \leq k} \left(\frac{E_{i_j}}{E_{i_{k+1}}} \right)^{\alpha_j} \right) E_{i_{k+1}}^{k+1} \leq \left(\prod_{1 \leq j \leq k} \frac{E_{i_j}}{E_{i_{k+1}}} \right) E_{i_{k+1}}^{k+1} = \prod_{1 \leq j \leq k+1} E_{i_j}.$$

The combination of the last two inequalities completes the proof of (v). □

Theorem 4.3.4 (Degenerated SMT). *Let $f = [f_0 : \cdots : f_n] : D(R_0) \to \mathbb{P}^n(\mathbb{C})$ be a holomorphic map whose image is contained in some k-dimensional subspace but not in any subspace of dimension lower than k, where $0 < R_0 \le \infty$. Let H_j, $1 \le j \le q$, be hyperplanes in $\mathbb{P}^n(\mathbb{C})$ located in general position. Assume that $f(D(R_0)) \not\subset H_j$ for $1 \le j \le q$. Then for every $\epsilon > 0$, we have*

$$\sum_{j=1}^{q} m_{f,H_j}(r) + \left(\frac{n+1}{k+1}\right) N(R_f, r) \le (2n-k+1)T_f(r) + \left(\frac{2n-k+1}{k+1}\right) S(r) + \left(\frac{k(2n-k+1)}{2}\right)\epsilon \log r$$

for $0 < r < R_0$, where $N(R_f, r)$ is the ramification term defined in the proof, and $S(r)$ is the same as in Theorem 4.2.3.

Proof. Without loss of generality, we may assume that $f(D(R_0)) \subset \mathbb{P}^k(\mathbb{C})$. So $f : D(R_0) \to \mathbb{P}^k(\mathbb{C})$ is a linearly nondegenerate holomorphic map. We also assume that $q \ge 2n-k+1$. Denote $\hat{H}_j = H_j \cap \mathbb{P}^k(\mathbb{C})$. Then $\hat{H}_j$ are hyperplanes in $\mathbb{P}^k(\mathbb{C})$ located in n-subgeneral position.

Recall that the Weil function of f is defined as

$$\lambda_H(f(z)) = \log \frac{\|\mathbf{f}(z)\|\|\mathbf{a}\|}{|\langle \mathbf{f}(z), \mathbf{a}\rangle|} \ge 0.$$

Since $H_1, \ldots, H_q$ are in general position in $\mathbb{P}^n(\mathbb{C})$, for each $z \in \mathbb{C}$, there exist indices $i(z,0), \ldots, i(z,n) \in \{1, \ldots, q\}$ such that

$$\sum_{j=1}^{q} \omega(j)\lambda_{H_j}(f(z)) \le \sum_{l=0}^{n} \omega(i(z,l))\lambda_{H_{i(z,l)}}(f(z)). \tag{4.52}$$

By Theorem 4.3.3 with $E_l = e^{\lambda_{\hat{H}_{i(z,l)}}(f(z))}$, $0 \le l \le n$, there is a subset M of $Y = \{i(z,0), \ldots, i(z,n)\}$ with $\#M = k+1$ such that $\{\hat{H}_{i(z,j)} \mid i(z,j) \in M\}$ is linearly independent and

$$\prod_{l=0}^{n} e^{\omega(i(z,l))\lambda_{\hat{H}_{i(z,l)}}(f(z))} \le \prod_{i(z,j)\in M} e^{\lambda_{\hat{H}_{i(z,j)}}(f(z))}.$$

Thus

$$\sum_{l=0}^{n} \omega(i(z,l))\lambda_{\hat{H}_{i(z,l)}}(f(z)) \le \sum_{i(z,j)\in M} \lambda_{\hat{H}_{i(z,j)}}(f(z)) \le \max_{\gamma\in\Gamma} \sum_{l=0}^{k} \lambda_{\hat{H}_{\gamma(l)}}(f(z)),$$

where Γ is the set of all maps $\gamma : \{0, \ldots, k\} \to \{1, \ldots, q\}$ such that $\hat{H}_{\gamma(0)}, \ldots, \hat{H}_{\gamma(k)}$ are linearly independent. Combining this with (4.52) gives us

$$\sum_{j=1}^{q}\omega(j)m_f(H_j,r)\le\int_0^{2\pi}\max_{\gamma\in\Gamma}\sum_{l=0}^{k}\lambda_{\hat{H}_{\gamma(l)}}(f(re^{i\theta}))\frac{d\theta}{2\pi}+O(1).$$

Applying Theorem 4.2.6 yields

$$\int_0^{2\pi}\max_{\gamma\in\Gamma}\sum_{l=0}^{k}\lambda_{\hat{H}_{\gamma(l)}}(f(re^{i\theta}))\frac{d\theta}{2\pi}+N(R_f,r)\le(k+1)T_f(r)+S(r)+\frac{k(k+1)}{2}\epsilon\log r,$$

where $N(R_f,r):=N_W(r,0)$. Hence

$$\sum_{j=1}^{q}\omega(j)m_f(H_j,r)+N(R_f,r)\le(k+1)T_f(r)+S(r)+\frac{k(k+1)}{2}\epsilon\log r.$$

By Theorem 4.3.3, recalling that $m_f(r,H_j)\le T_f(r)+O(1)$, this inequality gives

$$\begin{aligned}\sum_{j=1}^{q}m_f(r,H_j)&=\sum_{j=1}^{q}(1-\theta\omega(j))m_f(r,H_j)+\sum_{j=1}^{q}\theta\omega(j)m_f(r,H_j)\\&\le\sum_{j=1}^{q}(1-\theta\omega(j))m_f(r,H_j)+\theta(k+1)T_f(r)-\theta N(R_f,r)+\theta S(r)\\&\quad+\theta\frac{k(k+1)}{2}\epsilon\log r\\&\le\sum_{j=1}^{q}(1-\theta\omega(j))T_f(r)+\theta(k+1)T_f(r)-\left(\frac{n+1}{k+1}\right)N(R_f,r)+\theta S(r)\\&\quad+\theta\frac{k(k+1)}{2}\epsilon\log r\\&=\left\{q-\theta\left(\sum_{1\le j\le q}\omega(j)-k-1\right)\right\}T_f(r)-\left(\frac{n+1}{k+1}\right)N(R_f,r)+\theta S(r)\\&\quad+\theta\frac{k(k+1)}{2}\epsilon\log r\\&=(2n-k+1)T_f(r)-\left(\frac{n+1}{k+1}\right)N(R_f,r)+\left(\frac{2n-k+1}{k+1}\right)S(r)\\&\quad+\left(\frac{k(2n-k+1)}{2}\right)\epsilon\log r.\end{aligned}$$

□

Corollary 4.3.5. *Let $H_j, 1\le j\le q$, be hyperplanes in $\mathbb{P}^n(\mathbb{C})$ in general position, and let $q\ge 2n+1$. Then every holomorphic map $f:\mathbb{C}\to\mathbb{P}^n(\mathbb{C})-\bigcup_{j=1}^{q}H_j$ is constant.*

Proof. Assume that $f:\mathbb{C}\to\mathbb{P}^n(\mathbb{C})-\bigcup_{j=1}^{q}H_j$ is not constant. Then the image of f is contained in some k-dimensional subspace but not in any subspace of dimension lower than k with $1\le k\le n$. Applying Theorem 4.3.4 with $R_0=\infty$, we get

$$\sum_{j=1}^{q} m_{f,H_j}(r) \leq_{\text{exc}} (2n-k+1)T_f(r) + O(\log^+ T_f(r)) + \epsilon \log r.$$

Since the image of f omits H_j, we have $N_f(r, H_j) = 0$, so by the first main theorem, $m_f(r, H_j) = T_f(r) + O(1)$ for $1 \leq j \leq 2n+1$. Hence we get

$$(2n+1)T_f(r) \leq_{\text{exc}} (2n-k+1)T_f(r) + O(\log^+ T_f(r)) + \epsilon \log r,$$

that is, $kT_f(r) \leq_{\text{exc}} O(\log^+ T_f(r)) + \epsilon \log r$, which gives a contradiction. □

4.4 Holomorphic curves through Ahlfors' negative curved method

We follow the method by Ahlfors to extend the theory of algebraic curves (see Chapter 2) to holomorphic curves.

The associated maps

Let $D(R) = \{z \mid |z| < R\} \subset \mathbb{C}$ with $0 < R \leq +\infty$, and let $f : D(R) \to \mathbb{P}^n(\mathbb{C})$ be a linearly non-degenerate map, that is, the image of f is not contained in any proper subspace of $\mathbb{P}^n(\mathbb{C})$. Let $\mathbf{f} : D(R) \to \mathbb{C}^{n+1} - \{0\}$ be a reduced representation of f. Consider the holomorphic map

$$\mathbf{F}_k = \mathbf{f} \wedge \mathbf{f}' \wedge \cdots \wedge \mathbf{f}^{(k)} : D_R \to \bigwedge^{k+1} \mathbb{C}^{n+1}.$$

Evidently, $\mathbf{F}_{n+1} \equiv 0$. Throughout this section, we assume that f is linearly nondegenerate, so $\mathbf{F}_k \not\equiv 0$ for $0 \leq k \leq n$. The map $F_k = \mathbb{P}(\mathbf{F}_k) : D(R) \to \mathbb{P}(\bigwedge^{k+1} \mathbb{C}^{n+1}) = \mathbb{P}^{N_k}$, where $N_k = \frac{(n+1)!}{(k+1)!(n-k)!} - 1$ and $\mathbb{P}$ is the natural projection, is called *the kth associated map*. We can express them by using the Wronskians. For holomorphic functions $f_0, f_1, \ldots, f_n$, we define the Wronskian $W(f_0, f_1, \ldots, f_n)$ by

$$W(f_0, f_1, \ldots, f_n) := \det(f_j^{(s)}, 0 \leq j, s \leq n).$$

Let $\{e_0, e_1, \ldots, e_n\}$ be the standard basis of $\mathbb{C}^{n+1}$. Let $\mathbf{f} = (f_0, \ldots, f_n)$ be a reduced representation of f. Then

$$\mathbf{F}_k = \sum_{0 \leq j_0 < \cdots < j_k \leq n} W(f_{j_0}, f_{j_1}, \ldots, f_{j_k}) \mathbf{e}_{j_0} \wedge \cdots \wedge \mathbf{e}_{j_k}, \tag{4.53}$$

$$\|\mathbf{F}_k\|^2 = \sum_{0 \leq j_0 < \cdots < j_k \leq n} |W(f_{j_0}, f_{j_1}, \ldots, f_{j_k})|^2. \tag{4.54}$$

From the above we easily see that if f is linearly nondegenerate, then $\mathbf{F}_k \not\equiv 0$ for any $0 \leq k \leq n$.

Let $\omega_k = dd^c \log \|Z\|^2$ be the Fubini–Study form on $\mathbb{P}^{N_k}(\mathbb{C})$, where $Z = [x_0 : \cdots : x_{N_k}] \in \mathbb{P}^{N_k}(\mathbb{C})$. Let

$$\Omega_k = F_k^* \omega_k = \frac{\sqrt{-1}}{2\pi} h_k dz \wedge d\bar{z}, \quad 0 \le k \le n, \tag{4.55}$$

be the pullback via the kth associated curve. By Corollary 4.1.2

$$h_k = \frac{\|\mathbf{F}_{k-1}\|^2 \|\mathbf{F}_{k+1}\|^2}{\|\mathbf{F}_k\|^4} \tag{4.56}$$

for $0 \le k \le n$, and by convention $\|\mathbf{F}_{-1}\| \equiv 1$. Note that $\Omega_n \equiv 0$. It follows that

$$dd^c \log h_k = \Omega_{k-1} + \Omega_{k+1} - 2\Omega_k.$$

The projective distance

For integer $1 \le q \le p \le n+1$, the *interior product* $\xi \lfloor \alpha$ of vectors $\xi \in \bigwedge^{p+1} \mathbb{C}^{n+1}$ and $\alpha \in \bigwedge^{q+1} \mathbb{C}^{*n+1}$ is defined by

$$(\xi \lfloor \alpha, \beta) = (\xi, \alpha \wedge \beta) = (\alpha \wedge \beta)(\xi)$$

for any $\beta \in \bigwedge^{p-q} \mathbb{C}^{*n+1}$, where $(\cdot, \cdot)$ is the inner product. For a hyperplane H in $\mathbb{P}^n(\mathbb{C})$ with unit normal vector $\mathbf{a} \in \mathbb{C}^{*n+1}$ and $x \in \mathbb{P}(\bigwedge^{k+1} \mathbb{C}^{n+1})$, the *projective distance* is defined by

$$\|x; H\| = \frac{\|\xi \lfloor \mathbf{a}\|}{\|\xi\| \|\mathbf{a}\|}, \tag{4.57}$$

where $\xi \in \bigwedge^{k+1} \mathbb{C}^{n+1}$ with $\mathbb{P}(\xi) = x$. For the hyperplane $H = \{a_0 z_0 + \cdots + a_n z_n = 0\} \subset \mathbb{P}^n(\mathbb{C})$ with $\mathbf{a} = (a_0, \ldots, a_n)$, by using (4.53) we have

$$\|\mathbf{F}_k \lfloor \mathbf{a}\|^2 = \sum_{0 \le j_1 < \cdots < j_k \le n} \left| \sum_{j_0 \ne j_1, \ldots, j_k} a_{j_0} W(f_{j_0}, f_{j_1}, \ldots, f_{j_k}) \right|^2. \tag{4.58}$$

From the above we see that $\mathbf{F}_k \lfloor \mathbf{a} = \mathbf{0}$ if and only if

$$\sum_{t \ne i_1, \ldots, i_k} a_t W(f_t, f_{i_1}, \ldots, f_{i_k}) \equiv 0$$

for all $i_1, \ldots, i_k$. Thus if f is linearly nondegenerate, then $\mathbf{F}_k \lfloor \mathbf{a} \not\equiv \mathbf{0}$ for all $0 \le k \le n$. Indeed, if $\mathbf{F}_k \lfloor \mathbf{a} \equiv \mathbf{0}$ for some k, then

$$W(\mathbf{f}(H), f_{i_1}, \ldots, f_{i_k}) \equiv 0$$

for all $i_1, \ldots, i_k$, where $\mathbf{f}(H) := a_0 f_0 + \cdots + a_n f_n$. This implies that $\mathbf{f}(H), f_{i_1}, \ldots, f_{i_k}$ are linearly dependent, which contradicts the nondegeneracy assumption of f.

Some computations

Let

$$\phi_k(H) := \|F_k; H\|^2 = \frac{\|\mathbf{F}_k \lfloor \mathbf{a}\|^2}{\|\mathbf{F}_k\|^2\|\mathbf{a}\|^2}, \tag{4.59}$$

where $\mathbf{F}_k$ is a reduced representation of F_k, and H is the hyperplane defined by $\mathbf{a}$. Define

$$h_k(H) = \frac{\phi_{k-1}(H)\phi_{k+1}(H)}{\phi_k^2(H)}\Omega_k.$$

Lemma 4.4.1. *Let H be a hyperplane in $\mathbb{P}^n(\mathbb{C})$. For $0 \le k \le n$, we have*
(i) $dd^c \log \phi_k(H) = h_k(H) - \Omega_k$,
(ii) $0 \le \phi_{k-1}(H) \le \phi_k(H) \le \phi_n(H) = 1$,
(iii) $d\phi_k(H) \wedge d^c\phi_k(H) = (\phi_{k+1}(H) - \phi_k(H))(\phi_k(H) - \phi_{k-1}(H))\Omega_k$,
(iv) $dd^c\phi_k(H) = (\phi_{k+1}(H) - 2\phi_k(H) + \phi_{k-1}(H))\Omega_k$.

Proof. Since

$$\phi_k(H) = \frac{\|\mathbf{F}_k \lfloor \mathbf{a}\|^2}{\|\mathbf{F}_k\|^2\|a\|^2}$$

with unit vector $\mathbf{a}$,

$$dd^c \log \phi_k(H) = dd^c \log\|\mathbf{F}_k \lfloor \mathbf{a}\|^2 - dd^c \log \|\mathbf{F}_k\|^2 = dd^c \log\|\mathbf{F}_k \lfloor \mathbf{a}\|^2 - \Omega_k.$$

Now by (4.56)

$$\begin{aligned}
dd^c \log\|\mathbf{F}_k \lfloor \mathbf{a}\|^2 &= dd^c \log \|(\mathbf{F}_1 \lfloor \mathbf{a})_{k-1}\|^2 \\
&= \frac{\|(\mathbf{F}_1 \lfloor \mathbf{a})_{k-2}\|^2\|(\mathbf{F}_1 \lfloor \mathbf{a})_k\|^2}{\|(\mathbf{F}_1 \lfloor \mathbf{a})_{k-1}\|^4}\frac{\sqrt{-1}}{2\pi}dz \wedge d\bar{z} \\
&= \frac{\|\mathbf{F}_{k-1} \lfloor \mathbf{a}\|^2\|\mathbf{F}_{k+1} \lfloor \mathbf{a}\|^2}{\|\mathbf{F}_k \lfloor \mathbf{a}\|^4}\frac{\sqrt{-1}}{2\pi}dz \wedge d\bar{z} \\
&= \frac{\phi_{k-1}(H)\phi_{k+1}(H)}{\phi_k^2(H)}\frac{\|\mathbf{F}_{k-1}\|^2\|\mathbf{F}_{k+1}\|^2}{\|\mathbf{F}_k\|^4}\frac{\sqrt{-1}}{2\pi}dz \wedge d\bar{z} \\
&= \frac{\phi_{k-1}(H)\phi_{k+1}(H)}{\phi_k^2(H)}\Omega_k.
\end{aligned}$$

So (i) is proved by combining the above two identities.

We now prove (ii). For any $z \in \mathbb{C}$ with $\mathbf{F}_n(z) \ne 0$, by the Gram–Schmidt normalization process there exist numbers A_{ji} for $0 \le i \le j \le n$ with $A_{ii} \ne 0$ and an orthonormal basis $\mathbf{e}_0, \dots, \mathbf{e}_n$ of $\mathbb{C}^{n+1}$ such that

$$\mathbf{f}^{(j)}(z) = \sum_{i=0}^{j} A_{ji}\mathbf{e}_i \quad \text{for } j = 0, \ldots, n.$$

So

$$\mathbf{F}_j = A_{00} \cdots A_{jj}\mathbf{e}_0 \wedge \cdots \wedge \mathbf{e}_j \quad \text{for } j = 0, \ldots, n, \tag{4.60}$$

and

$$\|\mathbf{F}_j\| = |A_{00} \cdots A_{jj}| \quad \text{for } j = 0, \ldots, n. \tag{4.61}$$

Let $\mathbf{a} \in \mathbf{C}^{*^{n+1}}$ be the unit normal vector associated with the hyperplane H. Define $B_j = \mathbf{a}(\mathbf{e}_j)$ and $\mathbf{e}_q^j = (-1)^j \mathbf{e}_0 \wedge \cdots \wedge \mathbf{e}_{j-1} \wedge \mathbf{e}_{j+1} \wedge \cdots \wedge \mathbf{e}_q$ for $0 \leq j \leq q \leq n$. Note that

$$(\mathbf{e}_0 \wedge \cdots \wedge \mathbf{e}_q) \llcorner \mathbf{a} = \sum_{j=0}^{q} \mathbf{a}(\mathbf{e}_j)\mathbf{e}_q^j = \sum_{j=0}^{q} B_j \mathbf{e}_q^j. \tag{4.62}$$

So we have

$$\mathbf{F}_q \llcorner \mathbf{a} = A_{00} \cdots A_{qq} \sum_{j=0}^{q} B_j \mathbf{e}_q^j. \tag{4.63}$$

Thus by (4.61)

$$\|\mathbf{F}_q \llcorner \mathbf{a}\|^2 = \|\mathbf{F}_q\|^2 \sum_{j=0}^{q} |B_j|^2. \tag{4.64}$$

Hence

$$\phi_q(H) = \sum_{j=0}^{q} |B_j|^2, \quad \phi_q(H) - \phi_{q-1}(H) = |B_q|^2 \geq 0. \tag{4.65}$$

This gives $0 \leq \phi_{q-1}(H) \leq \phi_q(H)$ for $q = 1, \ldots, n$. Also,

$$\phi_n(H) = \sum_{j=0}^{n} |B_j|^2 = \sum_{j=0}^{n} |\mathbf{a}(\mathbf{e}_j)|^2 = \|\mathbf{a}\|^2 = 1.$$

Thus (ii) is proved.

We proceed to verify (iii). First of all,

$$\partial\phi_k(H) = \left(\frac{(\mathbf{F}_k' \llcorner \mathbf{a}, \mathbf{F}_k \llcorner \mathbf{a})}{\|\mathbf{F}_k\|^2} - \frac{(\mathbf{F}_k', \mathbf{F}_k)\|\mathbf{F}_k \llcorner \mathbf{a}\|^2}{\|\mathbf{F}_k\|^4} \right) dz.$$

We now calculate each individual term that appears in the above inequality:

$$\begin{aligned}\mathbf{F}_k' &= \mathbf{f}\wedge\mathbf{f}'\wedge\cdots\wedge\mathbf{f}^{(k-1)}\wedge\mathbf{f}^{(k+1)}\\ &= \left(\prod_{q=0}^{k-1}A_{qq}\right)\mathbf{e}_0\wedge\cdots\wedge\mathbf{e}_{k-1}\wedge(A_{k+1,k}\mathbf{e}_k+A_{k+1,k+1}\mathbf{e}_{k+1}).\end{aligned} \tag{4.66}$$

Using (4.62) and (4.66), we have

$$\begin{aligned}\mathbf{F}_k'\lfloor\mathbf{a} &= \left(\prod_{q=0}^{k-1}A_{qq}\right)\sum_{j=0}^{k-1}B_j\mathbf{e}_{k-1}^j\wedge(A_{k+1,k}\mathbf{e}_k+A_{k+1,k+1}\mathbf{e}_{k+1})\\ &\quad+\left(\prod_{q=0}^{k-1}A_{qq}\right)(B_kA_{k+1,k}+B_{k+1}A_{k+1,k+1})\mathbf{e}_k^k.\end{aligned}$$

Combining this identity with (4.63), we get

$$(\mathbf{F}_k'\lfloor\mathbf{a},\mathbf{F}_k\lfloor\mathbf{a})=\left(\prod_{q=0}^{k-1}|A_{qq}|^2\right)G,$$

where

$$G=\bar{A}_{k,k}\left(\sum_{j=0}^{k}|B_j|^2A_{k+1,k}+B_{k+1}A_{k+1,k+1}\bar{B}_k\right).$$

Using (4.66) and (4.60), we have

$$(\mathbf{F}_k',\mathbf{F}_k)=\left(\prod_{q=0}^{k-1}|A_{qq}|^2\right)\bar{A}_{k,k}A_{k+1,k}.$$

This, together with (4.64), gives

$$(\mathbf{F}_k',\mathbf{F}_k)\frac{\|\mathbf{F}_k\lfloor\mathbf{a}\|^2}{\|\mathbf{F}_k\|^2}=\left(\sum_{j=0}^{k}|B_j|^2\right)\left(\prod_{q=0}^{k-1}|A_{qq}|^2\right)\bar{A}_{k,k}A_{k+1,k}.$$

Therefore

$$\partial\phi_k(H)=\frac{A_{k+1,k+1}}{A_{kk}}B_{k+1}\overline{B}_kdz.$$

By the above identity, (4.65), and (4.56) we obtain

$$\begin{aligned}d\phi_k(H)\wedge d^c\phi_k(H) &= \frac{\sqrt{-1}}{2\pi}\partial\phi_k(H)\wedge\bar{\partial}\phi_k(H)\\ &= \frac{\sqrt{-1}}{2\pi}\frac{|A_{k+1,k+1}|^2}{|A_{kk}|^2}|B_{k+1}|^2|B_k|^2dz\wedge d\bar{z}\end{aligned}$$

$$= \frac{\sqrt{-1}}{2\pi} \frac{\|\mathbf{F}_{k-1}\|^2 \|\mathbf{F}_{k+1}\|^2}{\|\mathbf{F}_k\|^4} |B_{k+1}|^2 |B_k|^2 dz \wedge d\bar{z}$$
$$= (\phi_{k+1}(H) - \phi_k(H))(\phi_k(H) - \phi_{k-1}(H))\Omega_k.$$

Hence (iii) follows.

Also, using (i) and (iii), we have

$$\begin{aligned} dd^c \phi_k(H) &= \phi_k(H) dd^c \log \phi_k(H) + \frac{1}{\phi_k(H)} d\phi_k(H) \wedge d^c \phi_k(H) \\ &= \phi_k(H)(h_k(H) - \Omega_k) \\ &\quad + \frac{1}{\phi_k(H)}(\phi_{k+1}(H) - \phi_k(H))(\phi_k(H) - \phi_{k-1}(H))\Omega_k \\ &= \frac{1}{\phi_k(H)}(\phi_{k-1}(H)\phi_{k+1}(H) - \phi_k^2(H))\Omega_k \\ &\quad + \frac{1}{\phi_k(H)}(\phi_{k+1}(H) - \phi_k(H))(\phi_k(H) - \phi_{k-1}(H))\Omega_k \\ &= (\phi_{k+1}(H) - 2\phi_k(H) + \phi_{k-1}(H))\Omega_k. \end{aligned}$$

Thus (iv) is verified. □

Theorem 4.4.2. *For every $\epsilon > 0$, there exists a number $\mu_0(\epsilon) \geq 1$ such that for all $\mu \geq \mu_0(\epsilon)$ and for any hyperplane $H \subset \mathbb{P}^n(\mathbb{C})$, we have*

$$dd^c \log \frac{1}{\log^2(\mu/\phi_k(H))} \geq \frac{2\phi_{k+1}(H)}{\phi_k(H)\log^2(\mu/\phi_k(H))}\Omega_k - \epsilon\Omega_k$$

on $\{z \mid \phi_k(H)(z) \neq 0\}$.

Proof. First of all,

$$dd^c \log \frac{1}{\log^2(\mu/\phi_k(H))} = 2\frac{dd^c \log \phi_k(H)}{\log(\mu/\phi_k(H))} + 2\frac{d\phi_k(H) \wedge d^c \phi_k(H)}{\phi_k^2(H)\log^2(\mu/\phi_k(H))}. \tag{4.67}$$

Recall that by (i) and (iii) in Lemma 4.4.1 we have

$$dd^c \log \phi_k(H) = \frac{\phi_{k-1}(H)\phi_{k+1}(H) - \phi_k^2(H)}{\phi_k^2(H)}\Omega_k$$

and

$$d\phi_k(H) \wedge d^c \phi_k(H) = (\phi_{k+1}(H) - \phi_k(H))(\phi_k(H) - \phi_{k-1}(H))\Omega_k.$$

Thus we can rewrite (4.67) in the form

$$
\begin{aligned}
dd^c \log \frac{1}{\log^2(\mu/\phi_k(H))} &= \frac{2\phi_{k+1}(H)}{\phi_k(H)\log^2(\mu/\phi_k(H))}\Omega_k \\
&\quad + 2\frac{\phi_{k-1}(H)\phi_{k+1}(H)}{\phi_k^2(H)}\left(\frac{1}{\log(\mu/\phi_k(H))} - \frac{1}{\log^2(\mu/\phi_k(H))}\right)\Omega_k \\
&\quad + 2\frac{\phi_{k-1}(H)}{\phi_k(H)\log^2(\mu/\phi_k(H))}\Omega_k \\
&\quad - 2\left(\frac{1}{\log(\mu/\phi_k(H))} + \frac{1}{\log^2(\mu/\phi_k(H))}\right)\Omega_k.
\end{aligned}
$$

Then since $\phi_k(H) \leq 1$, we can choose a number μ so large that $\log(\mu/\phi_k(H)) > 1$ and

$$\frac{1}{\log(\mu/\phi_k(H))} + \frac{1}{\log^2(\mu/\phi_k(H))} < \frac{\epsilon}{2}$$

for a given $\epsilon > 0$ and

$$\frac{1}{\log(\mu/\phi_k(H))} - \frac{1}{\log^2(\mu/\phi_k(H))} > 0.$$

Then

$$dd^c \log \frac{1}{\log^2(\mu/\phi_k(H))} \geq \frac{2\phi_{k+1}(H)}{\phi_k(H)\log^2(\mu/\phi_k(H))}\Omega_k - \epsilon\Omega_k. \qquad \square$$

The metric with negative curvature

Let $f : D(R_0) \to \mathbb{P}^m(\mathbb{C})$, $0 < R_0 \leq \infty$, be a holomorphic map. We assume that f is n-linearly nondegenerate, that is, $f(D(R_0))$ is contained in a subspace of dimension $n \leq m$ but not in any subspace of lower dimension. Without a loss of generality, we assume that $f : D(R_0) \to \mathbb{P}^n(\mathbb{C})$. Then f is linearly nondegenerate. Furthermore, the hyperplanes $H_j \cap \mathbb{P}^n(\mathbb{C}), 1 \leq j \leq q$, are in m-subgeneral position. Hence, in this section, we consider a linearly nondegenerate holomorphic map $f : D(R_0) \to \mathbb{P}^n(\mathbb{C})$ with $0 < R_0 \leq \infty$ and a set of hyperplanes $H_1, \dots, H_q$ in $\mathbb{P}^n(\mathbb{C})$ located in m-subgeneral position with $m \geq n$. Let $\omega(j)$ be the Nochka weights associated with H_j for $1 \leq j \leq q$.

The following product-to-sum estimate is an extension of the estimate of the geometric mean by the arithmetic mean. First, we prove the following lemma.

Lemma 4.4.3. *Let $H_1, \dots, H_q$ (or $\mathbf{a}_1, \dots, \mathbf{a}_q$) be hyperplanes in $\mathbb{P}^n(\mathbb{C})$ in m-subgeneral position. Let $p \in \mathbb{Z}[0, n]$ and $x \in \mathbb{P}(\bigwedge^p \mathbb{C}^{n+1})$. Define*

$$I_x = \{j \in \mathbb{N}[1, q] \mid \|x; H_j\| = 0\}.$$

Then $\#I_x \leq m - p$.

Proof. Let $k = \#I_x$, and let $E = \bigcap_{j\in I_x} H_j$. If $k \geq m+1$, then $E = \emptyset$ by the assumption of general position. If $k \leq m+1$, then $\dim E = m - k$. There are $\xi_j \in \mathbb{C}^{n+1}$ such that $0 \neq \xi := \xi_0 \wedge \cdots \wedge \xi_p$ and $x = \mathbb{P}(\xi)$. Let

$$E(x) = E(\xi) = \{\zeta \in \mathbb{C}^{n+1} \mid \zeta \wedge \xi = 0\}.$$

Then $E(x)$ is a linear subspace of dimension $p+1$ of $\mathbb{C}^{n+1}$ with base $\xi_0, \ldots, \xi_p$. Since $\|x; H_j\| = 0$ implies $\xi \lfloor \mathbf{a}_j = 0$ or $\langle \xi_\mu, \mathbf{a}_j \rangle = 0$ for $\mu = 0, \ldots, p$, we have $\mathbb{P}(E(x)) \subset E$, which implies

$$p = \dim \mathbb{P}(E(x)) \leq \dim E = m - k,$$

or $k \leq m - p$. □

Lemma 4.4.4. *Let $H_1, \ldots, H_q$ be hyperplanes in $\mathbb{P}^n(\mathbb{C})$ in m-subgeneral position. For $x \in \mathbb{P}(\bigwedge^p \mathbb{C}^{n+1})$ and a real number $d > 0$, define*

$$I_x(d) = \{j \in \mathbb{N}[1,q] \mid \|x; H_j\|^2 \leq d\}.$$

Then there is $c > 0$ such that $\#I_x(c) \leq m - p$ for all $x \in \mathbb{P}(\bigwedge^p \mathbb{C}^{n+1})$.

Proof. Take $x \in \mathbb{P}(\bigwedge^p \mathbb{C}^{n+1})$. Let I_x be the set defined in Lemma 4.4.3. There exists $c_x > 0$ such that $\|x; H_j\|^2 > c_x$ for all $j \in \mathbb{N}[1,q] - I_x$. There exists an open neighborhood U_x of x in $\mathbb{P}(\bigwedge^p \mathbb{C}^{n+1})$ such that $\|y; H_j\|^2 > c_x$ for all $y \in U_x$ and $j \in \mathbb{N}[1,q] - I_x$. Since $\mathbb{P}(\bigwedge^p \mathbb{C}^{n+1})$ is compact, there are only finitely many $U_{x_1}, \ldots, U_{x_s}$ whose union covers $\mathbb{P}(\bigwedge^p \mathbb{C}^{n+1})$. Define

$$c = \min\{c_{x_1}, \ldots, c_{x_s}\} > 0.$$

Take $y \in \mathbb{P}(\bigwedge^p \mathbb{C}^{n+1})$. Then there exists $\lambda \in \mathbb{N}[1,s]$ such that $y \in U_{x_\lambda}$. Take $j \in I_y(c)$. Assume that $j \notin I_{x_\lambda}$. Then

$$\|y; H_j\|^2 > c_{x_\lambda} \geq c \geq \|y; H_j\|^2,$$

which is impossible. Hence $I_y(c) \subset I_{x_\lambda}$ and $\#I_y(c) \leq \#I_{x_\lambda} \leq m - p$. □

Theorem 4.4.5 (Product-to-sum estimate). *Let $H_1, \ldots, H_q$ (or $\mathbf{a}_1, \ldots, \mathbf{a}_q$) be hyperplanes in $\mathbb{P}^n(\mathbb{C})$ in m-general position. Let $\omega(j), j = 1\ldots,q$, be the Nochka weights. For any $1/q \leq \lambda_k \leq 1/(n-k)$, there exists a constant $C_k > 0$, which depends only on k and the given hyperplanes, such that*

$$\prod_{j=1}^{q}\left[\left(\frac{\phi_{k+1}(H_j)}{\phi_k(H_j)}\right)^{\omega(j)} \frac{1}{\log(\mu/\phi_k(H_j))}\right]^{\lambda_k} \leq C_k \sum_{j=1}^{q} \frac{\phi_{k+1}(H_j)}{\phi_k(H_j)\log(\mu/\phi_k(H_j))}$$

on $D_{R_0} - \bigcup_{j=1}^{q}\{\phi_k(H_j) = 0\}$.

Proof. By Lemma 4.4.4 there exists a constant $c > 0$ such that $\#I_x(c) \le m-k$ for all $x \in \mathbb{P}(\bigwedge^k \mathbb{C}^{n+1})$. Fix $z \in \mathbb{C}$ and let $x = F_k(z)$. There exists a subset $A \subset \{1,\dots,q\}$ such that $I_x(c) \subset A$ with $\#A = m+1$. By Nochka's theorem there exists $B \subset A$ with $\#B = n+1$ such that $\mathbf{a}_j, j \in B$, are linearly independent and

$$\prod_{j\in A}\left(\frac{\phi_{k+1}(H_j)}{\phi_k(H_j)}\right)^{\omega(j)} \le \prod_{j\in B}\frac{\phi_{k+1}(H_j)}{\phi_k(H_j)}.$$

For $j \notin A$ (hence $j \notin I_x(c)$), $\phi_{k+1}(H_j)/\phi_k(H_j) \le 1/\phi_k(H_j) \le 1/c$, we get

$$\prod_{j=1}^{q}\left(\frac{\phi_{k+1}(H_j)}{\phi_k(H_j)}\right)^{\omega(j)} \le \frac{1}{c^{\sum_{j\notin A}}}\prod_{j\in A}\left(\frac{\phi_{k+1}(H_j)}{\phi_k(H_j)}\right)^{\omega(j)} \le \frac{1}{c^{\sum_{j\notin A}}}\prod_{j\in B}\frac{\phi_{k+1}(H_j)}{\phi_k(H_j)}.$$

Now denote

$$I_x(d,B) = \{j \in B \mid \|x;H_j\|^2 \le d\}.$$

Since $H_j, j \in B$, is in general position, by Lemma 4.4.4 (with $m = n$) we indeed get $\#(I_x(d,B)) \le n-k$ for some $d > 0$. Hence there exists a subset $I \subset B$ such that $\#I = n-k$. Thus $\#(B-I) = n+1-(n-k) = k+1$, and for $j \in B-I$, $\phi_{k+1}(H_j)/\phi_k(H_j) \le 1/\phi_k(H_j) \le 1/d$, so we get

$$\begin{aligned}\prod_{j=1}^{q}\left(\frac{\phi_{k+1}(H_j)}{\phi_k(H_j)}\right)^{\omega(j)} &\le \frac{1}{c^{\sum_{j\notin A}}}\prod_{j\in B}\frac{\phi_{k+1}(H_j)}{\phi_k(H_j)}\\ &= \frac{1}{c^{\sum_{j\notin A}}}\prod_{j\in B-I}\frac{\phi_{k+1}(H_j)}{\phi_k(H_j)}\prod_{j\in I}\frac{\phi_{k+1}(H_j)}{\phi_k(H_j)}\\ &\le C\prod_{j\in I}\frac{\phi_{k+1}(H_j)}{\phi_k(H_j)}.\end{aligned}$$

Hence, noting that $1/(\log(\mu/\phi_k(H_j)) \le (1/\log\mu) := \tilde{c}$, we get

$$\begin{aligned}&\prod_{j=1}^{q}\left[\left(\frac{\phi_{k+1}(H_j)}{\phi_k(H_j)}\right)^{\omega(j)}\frac{1}{\log(\mu/\phi_k(H_j))}\right]\\ &\le \prod_{j=1}^{q}\left(\frac{\phi_{k+1}(H_j)}{\phi_k(H_j)}\right)^{\omega(j)}\prod_{j=1}^{q}\frac{1}{\log(\mu/\phi_k(H_j))}\\ &\le C\tilde{c}^{q-n+k}\prod_{j\in I}\frac{\phi_{k+1}(H_j)}{\phi_k(H_j)}\prod_{j\in I}\frac{1}{\log(\mu/\phi_k(H_j))}\\ &= \hat{C}\prod_{j\in I}\frac{\phi_{k+1}(H_j)}{\phi_k(H_j)\log(\mu/\phi_k(H_j))}.\end{aligned}$$

On the other hand, we have

$$\prod_{j\in I}\left(\frac{\phi_{k+1}(H_j)}{\phi_k(H_j)\log(\mu/\phi_k(H_j))}\right)^{1/(n-k)} \leq \frac{1}{n-k}\sum_{j\in I}\frac{\phi_{k+1}(H_j)}{\phi_k(H_j)\log(\mu/\phi_k(H_j))}$$
$$\leq \frac{1}{n-k}\sum_{j=1}^{q}\frac{\phi_{k+1}(H_j)}{\phi_k(H_j)\log(\mu/\phi_k(H_j))},$$

since the arithmetic mean majorizes the geometric mean. Thus

$$\prod_{j=1}^{q}\left[\left(\frac{\phi_{k+1}(H_j)}{\phi_k(H_j)}\right)^{\omega(j)}\frac{1}{\log(\mu/\phi_k(H_j))}\right]^{\lambda_k}$$
$$\leq \hat{C}^{\lambda_k}\prod_{j\in I}\left[\left(\frac{\phi_{k+1}(H_j)}{\phi_k(H_j)}\right)\frac{1}{\log(\mu/\phi_k(H_j))}\right]^{\lambda_k}$$
$$\leq \hat{C}^{\lambda_k}\prod_{j\in I}\left(\frac{\phi_{k+1}(H_j)}{\phi_k(H_j)\log(\mu/\phi_k(H_j))}\right)^{1/(n-k)}$$
$$\leq \frac{\hat{C}^{\lambda_k}}{n-k}\sum_{j=1}^{q}\frac{\phi_{k+1}(H_j)}{\phi_k(H_j)\log(\mu/\phi_k(H_j))},$$

and the lemma is proved. □

Let $f : D(R_0) \to \mathbb{P}^n(\mathbb{C})$ be a linearly nondegenerate holomorphic map with $R_0 \leq \infty$. Let $H_1, \ldots, H_q$ hyperplanes in $\mathbb{P}^n(\mathbb{C})$ located in m-subgeneral position. We use f to construct a pseudo-metric on $D(R_0)$. We write

$$\Omega_k = \frac{\sqrt{-1}}{2\pi}a_k(z)dz\wedge d\bar{z}, \tag{4.68}$$

where Ω_k is defined in (4.55). Let

$$\sigma_k = C_k^{-1}\prod_{j=1}^{q}\left[\left(\frac{\phi_{k+1}(H_j)}{\phi_k(H_j)(N-\log\phi_k(H_j))^2}\right)^{\omega(j)}\right]^{\lambda_k}\cdot a_k,$$

where C_k is the positive constant in the product-to-sum estimate above, $\lambda_k = 1/[n-k+2q(n-k)^2/N]$, and $N \geq 1$. We take the geometric mean of σ_k and define

$$\Gamma = \frac{\sqrt{-1}}{2\pi}c\prod_{k=0}^{n-1}\sigma_k^{\beta_n/\lambda_k}dz\wedge d\bar{z},$$

where $\beta_n = 1/\sum_{k=0}^{n-1}\lambda_k^{-1}$ and $c = 2(\prod_{k=0}^{n-1}\lambda_k^{\lambda_k^{-1}})^{\beta_n}$. Let

$$\Gamma = \frac{\sqrt{-1}}{2\pi}h(z)dz\wedge d\bar{z}.$$

Then

$$h(z) = \left(c\prod_{k=0}^{n-1} C_k^{-\frac{\beta_n}{\lambda_k}}\right)\prod_{j=1}^{q}\left(\frac{1}{\phi_0(H_j)^{\omega(j)}}\right)^{\beta_n}\prod_{j=1}^{q}\left[\prod_{k=0}^{n-1}\frac{a_k^{\frac{\beta_n}{q\lambda_k}}}{(N-\log\phi_k(H_j))^{2\beta_n}}\right]. \tag{4.69}$$

Theorem 4.4.6. *For $q \geq 2m - n + 2$ and*

$$2q/N < \frac{\sum_{j=1}^{q}\omega(j) - (n+1)}{n(n+2)},$$

we have

$$dd^c \log h(z) \geq \frac{\sqrt{-1}}{2\pi}h(z)dz \wedge d\bar{z}.$$

Proof. From (4.69) it follows that

$$dd^c \log h(z) = -\beta_n\sum_{j=1}^{q}\omega(j)dd^c\log\phi_0(H_j) + \beta_n\sum_{j=1}^{q}\sum_{k=0}^{n-1}dd^c\log\left(\frac{1}{N-\log\phi_k(H_j)}\right)^2$$
$$+\beta_n\sum_{k=1}^{n-1}(1/\lambda_k)dd^c\log a_k.$$

By (4.68) and (4.56), $dd^c\log a_k = \Omega_{k+1} - 2\Omega_k + \Omega_{k-1}$ and $dd^c\log\phi_0(H_j) = -\Omega_0$. These, together with Theorem 4.4.2, imply that

$$dd^c\log h(z) \geq \beta_n\left(\sum_{j=1}^{q}\omega(j)\Omega_0 + 2\sum_{j=1}^{q}\sum_{k=0}^{n-1}\frac{\phi_{k+1}(H_j)}{\phi_k(H_j)(N-\log\phi_k(H))^2}\Omega_k\right.$$
$$\left.-\frac{2q}{N}\sum_{k=0}^{n-1}\Omega_k + \sum_{k=0}^{n-1}\left[(n-k)+(n-k)^2\frac{2q}{N}\right]\{\Omega_{k+1} - 2\Omega_k + \Omega_{k-1}\}\right).$$

By Theorem 4.4.5 it follows that

$$\sum_{j=1}^{q}\frac{\phi_{k+1}(H_j)}{\phi_k(H_j)(N-\log\phi_k(H_j))^2}\Omega_k$$
$$\geq C_k^{-1}\left(\prod_{j=1}^{q}\left(\frac{\phi_{k+1}(H_j)}{\phi_k(H_j)}\right)^{\omega(j)}\frac{1}{(N-\log\phi_k(H_j))^2}\right)^{\lambda_k}\Omega_k$$
$$= \frac{\sqrt{-1}}{2\pi}\sigma_k(z)dz\wedge d\bar{z}.$$

Notice that $\Omega_n = 0$, so that

$$\sum_{k=0}^{n-1}(n-k)(\Omega_{k+1}-2\Omega_k+\Omega_{k-1})=-(n+1)\Omega_0,$$

and therefore

$$dd^c\log h(z)\ge\beta_n\left(\sum_{j=1}^{q}\omega(j)\Omega_0+2\frac{\sqrt{-1}}{2\pi}\sum_{k=0}^{n-1}\sigma_k(z)dz\wedge d\bar z-(n+1)\Omega_0-(n^2+2n)\frac{2q}{N}\Omega_0\right.$$
$$\left.+\sum_{k=1}^{n-2}[(n-k+1)^2-2(n-k)^2+(n-k-1)^2-1]\frac{2q}{N}\Omega_k+\frac{2q}{N}\Omega_{n-1}\right).$$

We use the following elementary inequality: For positive numbers $x_1,\dots,x_n$ and $a_1,\dots,a_q$,

$$a_1x_1+\cdots+a_nx_n\ge(a_1+\cdots+a_n)(x_1^{a_1}\cdots x_n^{a_n})^{1/(a_1+\cdots+a_n)}.$$

Letting $a_k=\lambda_k^{-1}$ and $x_k=\lambda_k\sigma_k$, we have

$$\sum_{k=0}^{n-1}\sigma_k\ge\frac{c}{2\beta_n}\prod_{k=0}^{n-1}\sigma_k^{\beta_n/\lambda_k}=\frac{h(z)}{2\beta_n},$$

and therefore

$$dd^c\log h(z)\ge\beta_n\left[\left(\sum_{j=1}^{q}\omega(j)-(n+1)-(n^2+2n)\frac{2q}{N}\right)\Omega_0+\sum_{k=1}^{n-2}\frac{2q}{N}\Omega_k\right.$$
$$\left.+\frac{2q}{N}\Omega_{n-1}\right]+\frac{\sqrt{-1}}{2\pi}h(z)dz\wedge d\bar z.$$

By Theorem 4.3.3 we have

$$\theta\left(\sum_{j=1}^{q}\omega(j)-(n+1)\right)=q-2m+n-1>0$$

and $\theta>0$, so $(\sum_{j=1}^{q}\omega(j)-(n+1))>0$. This and the choice of N give us

$$dd^c\log h(z)\ge\frac{\sqrt{-1}}{2\pi}h(z)dz\wedge d\bar z.\qquad\square$$

We use this alternative "negative curvature method" to obtain the following theorem again (see Corollary 4.3.5).

Theorem 4.4.7. *Let $H_j, 1\le j\le q$, be hyperplanes in $\mathbb{P}^n(\mathbb{C})$ in general position, and let $q\ge 2n+1$. Then every holomorphic map $f:\mathbb{C}\to\mathbb{P}^n(\mathbb{C})-\bigcup_{j=1}^{q}H_j$ is constant.*

Proof. Let h be defined above. By the Ahlfors–Schwarz lemma we have

$$h(z) \le \left(\frac{2R}{R^2 - |z|^2}\right)^2.$$

Letting $R \to \infty$, we have $h(z) \equiv 0$ on $\mathbb{C}$. Hence f is constant. □

A complex manifold M is said to be *Brody hyperbolic* if every holomorphic map $f : \mathbb{C} \to M$ is constant. From the above we know that $M := \mathbb{P}^n(\mathbb{C}) - \bigcup_{j=1}^q H_j$, where $H_1, \dots, H_q$ are in general position and $q \ge 2n + 1$, is Brody hyperbolic. Indeed, it is also complete Kobayashi hyperbolic and is hyperbolically imbedded in $\mathbb{P}^n(\mathbb{C})$. For the notations and details, see Lang [47] or Ru [76].

Theorem 4.4.8 (See Lang [47] or Ru [76]). *Let $H_1, \dots, H_q$ be hyperplanes in $\mathbb{P}^n(\mathbb{C})$, located in general position. If $q \ge 2n + 1$, then $X = \mathbb{P}^n(\mathbb{C}) - \bigcup_{j=1}^q H_j$ is complete hyperbolic and hyperbolically imbedded in $\mathbb{P}^n(\mathbb{C})$. Hence if $D \subset \mathbb{C}$ is the unit disc and Φ is a subset of* $\mathrm{Hol}(D, X)$, *then Φ is relatively locally compact in* $\mathrm{Hol}(D, \mathbb{P}^n(\mathbb{C}))$, *that is, given a sequence $\{f_n\}$ in Φ, there exists a subsequence that converges uniformly on every compact subset of D to an element of* $\mathrm{Hol}(D, \mathbb{P}^n(\mathbb{C}))$.

Denote $|\mathbf{F}_k(H)| := \|\mathbf{F}_k \lfloor \mathbf{a}\|$ whose formula is given in (4.58). As a consequence of Theorem 4.4.6, we get the following:

Lemma 4.4.9. *Let $F : D(R) \to \mathbb{P}^n(\mathbb{C})$ be a linearly nondegenerate holomorphic map. Let $\{H_j\}_{j=1}^q$ be a set of hyperplanes in $\mathbb{P}^n(\mathbb{C})$ in m-subgeneral position, and let $\omega(j)$ be their Nochka weights. Let $\mathbf{F}$ be a reduced representation of F. If $q > 2m - n + 1$ and $N > \frac{2qn(n+2)}{\sum_{j=1}^q \omega(j) - (n+1)}$, then there exists a constant $C > 0$ such that*

$$\left(\|\mathbf{F}\|^{\chi} \frac{\|\mathbf{F}_n\|^{1+\frac{2q}{N}} \prod_{j=1}^q \prod_{p=0}^{n-1} |\mathbf{F}_p(H_j)|^{\frac{4}{N}}}{\prod_{j=1}^q |\mathbf{F}(H_j)|^{\omega(j)}}\right)(z) \le C\left(\frac{2R}{R^2 - |z|^2}\right)^{\frac{1}{2}n(n+1) + \frac{2q}{N}\sum_{s=0}^n s^2},$$

where $\chi = \sum_{j=1}^q \omega(j) - (n + 1) - \frac{2q}{N}(n^2 + 2n - 1)$.

Proof. We will calculate $a_p^{\beta_n/\lambda_p}$ that appeared in (4.69). By (4.56)

$$a_p^{1/\lambda_p} = \left(\frac{\|\mathbf{F}_{p-1}\|^2 \|\mathbf{F}_{p+1}\|^2}{\|\mathbf{F}_p\|^4}\right)^{(n-p)+2q(n-p)^2/N}.$$

Therefore

$$\prod_{p=0}^{n-1} a_p^{1/\lambda_p} = \|\mathbf{F}_0\|^{-2(n+1)-(n^2+2n-1)4q/N} \|\mathbf{F}_1\|^{8q/N} \cdots \|\mathbf{F}_{n-1}\|^{8q/N} \|\mathbf{F}_n\|^{2+4q/N}.$$

Since $\mathbf{F}_0 = \mathbf{F}$ and $\phi_p(H) = \frac{|\mathbf{F}_p(H)|^2}{\|\mathbf{F}_p\|^2}$, by (4.69) and the inequality

$$h(z) \leq \left(\frac{2R}{R^2 - |z|^2}\right)^2$$

we get

$$\|\mathbf{F}(z)\|^{\chi} \frac{(\|\mathbf{F}_1(z)\| \cdots \|\mathbf{F}_{n-1}(z)\|)^{4q/N} \|\mathbf{F}_n(z)\|^{1+2q/N}}{\prod_{j=1}^{q}(|\mathbf{F}(H_j)(z)|^{\omega(j)} \prod_{p=0}^{n-1}(N - \log \phi_p(H_j)(z)))} \leq C\left(\frac{2R}{R^2 - |z|^2}\right)^{1/\beta_n}.$$

Set $K := \sup_{0<x\leq 1} x^{2/N}(N - \log x)$. Since $\phi_p(H_j) < 1$ for all p and j, we have

$$\frac{1}{(N - \log \phi_p(H_j))} \geq \frac{1}{K}\phi_p(H_j)^{2/N} = \frac{1}{K}\frac{|\mathbf{F}_p(H_j)|^{\frac{4}{N}}}{\|\mathbf{F}_p\|^{4/N}}.$$

Substituting these into the above inequality, we get the conclusion. □

5 Gauss maps of minimal surfaces in $\mathbb{R}^m$

In this chapter, we study the value distributions for the generalized Gauss maps of minimal surfaces immersed $\mathbb{R}^m$ using the theory of holomorphic curves developed in the previous chapter. This project was initiated by Chern and Osserman [20] in the mid-1960s. Related works include [24, 25, 28, 29, 43, 48, 66].

5.1 Minimal surfaces in $\mathbb{R}^m$

We extend the concept of mean curvature for surfaces in $\mathbb{R}^3$ introduced in Section 1.2 to surfaces in $\mathbb{R}^m$ with $m \geq 3$. Let M be a surface immersed in $\mathbb{R}^m$, and let $\mathbf{x} : M \to \mathbb{R}^m$ be an immersion. Take a local coordinate $\{U, \phi = (u^1, u^2)\}$ of M, where U is an open set of M. We denote $\mathbf{x}(u^1, u^2) := \mathbf{x} \circ \phi^{-1}(u^1, u^2)$. Then $\mathbf{x}(u^1, u^2) : U \to M$ is a local parameterization of M. Denote $\mathbf{x}_i := \frac{\partial \mathbf{x}}{\partial u^i}$, $i = 1, 2$. For $P \in M$, the tangent space of M at $P \in M$ is

$$T_P(M) = \{\lambda \mathbf{x}_1|_P + \mu \mathbf{x}_2|_P \mid \lambda, \mu \in \mathbb{R}\}.$$

The normal space of M at $P \in M$, that is, the space of all vectors normal to M at P, is defined by

$$N_P(M) := \{N \in \mathbb{R}^m \mid N \cdot \mathbf{x}_1|_P = N \cdot \mathbf{x}_2|_P = 0\}.$$

The metric, or the first fundamental form, is

$$g_{ij} = \mathbf{x}_i \cdot \mathbf{x}_j,$$

and for a fixed unit normal vector $N \in N_P(M)$, *the coefficients of the second fundamental form with respect to N* are

$$h_{ij,N} = \left(\frac{\partial^2 \mathbf{x}}{\partial u^i \partial u^j} \right) \cdot N, \quad 1 \leq i, j \leq 2.$$

Let $v \in T_P(M)$. The *normal curvature of M at P in the direction of v with respect to N* is

$$\kappa_{N,v} = \boldsymbol{\alpha}''(0) \cdot N,$$

where $\boldsymbol{\alpha}$ is a (unit-speed) curve $\boldsymbol{\alpha} : (-\epsilon, \epsilon) \to M$ with $\boldsymbol{\alpha}(0) = P$ and $\boldsymbol{\alpha}'(0) = v$. Set

$$\kappa_{1,N}(P) := \max_{v \in T_P(M), \|v\|=1} \kappa_{N,v}, \quad \kappa_{2,N}(P) := \min_{v \in T_P(M), \|v\|=1} \kappa_{N,v},$$

which are called the *principal curvatures of M at P with respect to N*. The *mean curvature of M at P with respect to N* is defined as

https://doi.org/10.1515/9783110989557-005

$$H_N(P) := \frac{\kappa_{1,N}(P) + \kappa_{2,N}(P)}{2}.$$

Similarly to the case in Section 1.2, we have the following formula for $P \in M$:

$$H_N(P) = \frac{g_{11}h_{22,N} + g_{22}h_{11,N} - 2g_{12}h_{12,N}}{2(g_{11}g_{22} - g_{12}^2)}. \tag{5.1}$$

Definition 5.1.1. A surface M immersed in $\mathbb{R}^m$ is called a *minimal surface* if $H_N(P) = 0$ for all $P \in M$ and $N \in N_P(M)$.

Similarly to Lemma 3.2.1, we have the following:

Lemma 5.1.2 (Existence of isothermal parameters on a minimal surface). *Let M be a minimal surface immersed in $\mathbb{R}^m$. Then there is a family of isothermal parameters that cover the totality of M.*

Proof. Since it is a local problem, we can have a local parameterization of $\mathbf{x}(x_1, x_2) = (x_1, x_2, f_3(x_1, x_2), \ldots, f_m(x_1, x_2)) : U \subset \mathbb{R}^2 \to \mathbb{R}^m$ for M. Consider the vector-valued function $f := (f_3, \ldots, f_m)$ and set

$$p := \frac{\partial f}{\partial x_1}, \quad q := \frac{\partial f}{\partial x_2}, \quad r := \frac{\partial^2 f}{\partial x_1^2}, \quad s := \frac{\partial^2 f}{\partial x_1 \partial x_2}, \quad t := \frac{\partial^2 f}{\partial x_2^2}.$$

We then have

$$g_{11} = 1 + \|p\|^2, \quad g_{12} = g_{21} = p \cdot q, \quad g_{22} = 1 + \|q\|^2.$$

The normal vectors $N = (N_1, \ldots, N_m)$ can be represented as

$$N_1 = -\sum_{k=3}^{m} N_k \frac{\partial f_k}{\partial x_1}, \quad N_2 = -\sum_{k=3}^{m} N_k \frac{\partial f_k}{\partial x_2}.$$

Notice that N is normal to the surface if and only if $N \cdot \mathbf{x}_1 = N \cdot \mathbf{x}_2$, so we see that $N_3, \ldots, N_m$ can be taken as arbitrary constants. By (5.1), M is minimal if and only if

$$g_{11}\frac{\partial^2 f}{\partial x_2^2} - 2g_{12}\frac{\partial^2 f}{\partial x_1 \partial x_2} + g_{22}\frac{\partial^2 f}{\partial x_1^2} = 0,$$

that is,

$$\Psi := (1 + \|q\|^2)r - 2(p \cdot q)s + (1 + \|p\|^2)t = 0.$$

Let $A = \sqrt{g_{11}g_{22} - g_{12}^2} = ((1 + \|p\|^2)(1 + \|q\|^2) - (p \cdot q)^2)^{1/2}$. Then we have

$$-\frac{\partial}{\partial x_1}\left(\frac{(p \cdot q)}{A}\right) + \frac{\partial}{\partial x_2}\left(\frac{1 + \|p\|^2}{A}\right) = \frac{1}{A^3}[(p \cdot q)p - (1 + \|p\|^2)q] \cdot \Psi.$$

Hence the minimality of M implies that

$$-\frac{\partial}{\partial x_1}\left(\frac{(p\cdot q)}{A}\right)+\frac{\partial}{\partial x_2}\left(\frac{1+\|p\|^2}{A}\right)=0,$$

that is,

$$\frac{\partial}{\partial x_1}\left(\frac{(p\cdot q)}{A}\right)=\frac{\partial}{\partial x_2}\left(\frac{1+\|p\|^2}{A}\right).$$

Similarly, we can show that

$$\frac{\partial}{\partial x_2}\left(\frac{(p\cdot q)}{A}\right)=\frac{\partial}{\partial x_1}\left(\frac{1+\|q\|^2}{A}\right).$$

Thus there are smooth functions ϕ and ψ on $U\subset\mathbb{R}^2$ such that

$$\phi_{x_1}=\frac{1+\|p\|^2}{A},\quad \phi_{x_2}=\psi_{x_1}=\frac{(p\cdot q)}{A},\quad \psi_{x_2}=\frac{1+\|q\|^2}{A}.$$

Set $L(x_1,x_2)=(x_1+\phi(x_1,x_2),x_2+\psi(x_1,x_2))=(u,v):U\subset\mathbb{R}^2\to\mathbb{R}^2$ (here L is called the Levy map). Then the Jacobi matrix of L is

$$J=\begin{pmatrix}1+\frac{1+\|p\|^2}{A} & \frac{(p\cdot q)}{A}\\ \frac{(p\cdot q)}{A} & 1+\frac{1+\|q\|^2}{A}\end{pmatrix},$$

and hence

$$\frac{\partial(u,v)}{\partial(x,y)}=2+\frac{\|p\|^2+\|q\|^2+2}{A}>0.$$

Therefore by the inverse function theorem the inverse L^{-1} exists (locally). Let $\tilde{\mathbf{x}}:=\mathbf{x}\circ L^{-1}$. We prove that this new parameterization is isothermal. Indeed, by the chain rule,

$$\begin{pmatrix}\tilde{\mathbf{x}}_u\\ \tilde{\mathbf{x}}_v\end{pmatrix}=\begin{pmatrix}(x_1)_u & (x_2)_u\\ (x_1)_v & (x_2)_v\end{pmatrix}\begin{pmatrix}\mathbf{x}_{x_1}\\ \mathbf{x}_{x_2}\end{pmatrix}.$$

Hence

$$\begin{pmatrix}\tilde{\mathbf{x}}_u\\ \tilde{\mathbf{x}}_v\end{pmatrix}^t\begin{pmatrix}\tilde{\mathbf{x}}_u\\ \tilde{\mathbf{x}}_v\end{pmatrix}=J^{-1}\begin{pmatrix}1+\|p\|^2 & p\cdot q\\ p\cdot q & 1+\|q\|^2\end{pmatrix}(J^{-1})^t=\frac{A}{\det J}\begin{pmatrix}1 & 0\\ 0 & 1\end{pmatrix}.$$

Hence we get

$$\tilde{\mathbf{x}}_u\cdot\tilde{\mathbf{x}}_u=\tilde{\mathbf{x}}_v\cdot\tilde{\mathbf{x}}_v=\frac{A}{\det J},\quad \tilde{\mathbf{x}}_u\cdot\tilde{\mathbf{x}}_v=0.$$

Therefore the new parameterization $\tilde{\mathbf{x}}$ is isothermal. □

Let M be a minimal surface immersed in $\mathbb{R}^m$. Similarly in Section 3.1, by Proposition 3.2.2, M becomes a Riemann surface with local isothermal coordinates (u,v) by letting $z = u + iv$. Let $\mathbf{x} : M \to \mathbb{R}^m$ be an immersion. Let $\{U, \phi = (u,v)\}$ be a local isothermal coordinate. We write $\mathbf{x}(u,v) = (\mathbf{x} \circ \phi^{-1})(u,v)$, which is a local parameterization for M. The induced metric on $M \subset \mathbb{R}^m$ from the standard Euclidean metric on $\mathbb{R}^m$ is

$$ds^2 = \lambda^2(du^2 + dv^2),$$

where $\lambda = \mathbf{x}_u \cdot \mathbf{x}_u = \mathbf{x}_v \cdot \mathbf{x}_v$. We claim that

$$\triangle_{ds^2} \mathbf{x} = 2HN, \tag{5.2}$$

where $\triangle_{ds^2}$ is the *Laplacian operator with respect to this induced metric* given by

$$\triangle_{ds^2} := \frac{1}{\lambda^2}\left(\frac{\partial}{\partial u^2} + \frac{\partial}{\partial v^2}\right) = \frac{4}{\lambda^2}\frac{\partial^2}{\partial z \partial \bar{z}}.$$

To prove this equation, notice that $\mathbf{x}(u,v)$ is isothermal, so $\mathbf{x}_u \cdot \mathbf{x}_u = \mathbf{x}_v \cdot \mathbf{x}_v$ and $\mathbf{x}_u \cdot \mathbf{x}_v = 0$. By differentiating these equations with respect to u and v we get

$$\mathbf{x}_{uu} \cdot \mathbf{x}_u = \mathbf{x}_{uv} \cdot \mathbf{x}_v \quad \text{and} \quad \mathbf{x}_{vu} \cdot \mathbf{x}_v = -\mathbf{x}_{vv} \cdot \mathbf{x}_u.$$

Hence

$$(\mathbf{x}_{uu} + \mathbf{x}_{vv}) \cdot \mathbf{x}_u = \mathbf{x}_{uv} \cdot \mathbf{x}_v - \mathbf{x}_{vu} \cdot \mathbf{x}_v = 0.$$

Similarly, we can get $(\mathbf{x}_{uu} + \mathbf{x}_{vv}) \cdot \mathbf{x}_v = 0$. From this we conclude that $\triangle_{ds^2}\mathbf{x}$ is proportional to N. On the other hand, for every normal vector,

$$H(N) = \frac{h_{11,N} + h_{22,N}}{2\lambda^2} = \frac{1}{2}(\triangle_{ds^2}\mathbf{x}) \cdot N.$$

This proves (5.2).

Therefore we get the following result.

Proposition 5.1.3. *A surface M in $\mathbb{R}^m$ is minimal if only if its isothermal parameterization $\mathbf{x}$ is harmonic.*

Write $\mathbf{x}(u,v) = (x_1(u,v), \dots, x_m(u,v))$. Let $\phi_k := \frac{\partial x_k}{\partial z} = \frac{1}{2}(\frac{\partial x_k}{\partial u} - i\frac{\partial x_k}{\partial v})$ for $k = 1, \dots, m$. Then the condition that M is minimal implies that $\mathbf{x}$ is harmonic, and hence $\frac{\partial^2 \mathbf{x}}{\partial z \partial \bar{z}} = \mathbf{0}$. Thus $\frac{\partial \phi_k}{\partial \bar{z}} = 0$, so ϕ_k is holomorphic for $k = 1, 2, \dots, m$. Also,

$$4\sum_{k=1}^{m} \phi_k^2 = \sum_{k=1}^{m}\left(\frac{\partial x_k}{\partial u}\right)^2 - \sum_{k=1}^{m}\left(\frac{\partial x_k}{\partial v}\right)^2 - 2i\sum_{k=1}^{m}\frac{\partial x_k}{\partial u}\frac{\partial x_k}{\partial v}$$

$$= \|\mathbf{x}_u\|^2 - \|\mathbf{x}_v\|^2 - 2i\mathbf{x}_u \cdot \mathbf{x}_v = 0$$

and

$$4\sum_{k=1}^{m} |\phi_k|^2 = \sum_{k=1}^{m}\left(\frac{\partial x_k}{\partial u}\right)^2 + \sum_{k=1}^{m}\left(\frac{\partial x_k}{\partial v}\right)^2 = \|\mathbf{x}_u\|^2 + \|\mathbf{x}_v\|^2 \neq 0$$

if M is regular. The differentials $\alpha_k := \phi_k dz$, $1 \leq k \leq 3$, are independent of the local coordinate z and hence are globally defined differential forms on M. Thus we obtain the following:

Theorem 5.1.4. *Let M be an open Riemann surface, and let $\omega_1, \omega_2, \ldots, \omega_m$ be holomorphic forms on M having no common zeros and no real periods and locally satisfying the identity*

$$f_1^2 + f_2^2 + \cdots + f_m^2 = 0$$

for holomorphic functions f_i with $\omega_i = f_i dz$. Set

$$x_i = 2\,\mathrm{Re}\int_{z_0}^{z} \omega_i$$

for an arbitrary fixed point z_0 of M. Then the surface $x = (x_1, \ldots, x_m) : M \to \mathbb{R}^m$ is a minimal surface immersed in $\mathbb{R}^m$.

Definition 5.1.5. The *generalized Gauss map* of $M \subset \mathbb{R}^m$ is defined as

$$G : M \to \mathbb{P}^{m-1}(\mathbb{C}), \quad G = \mathbb{P}\left(\frac{\partial \mathbf{x}}{\partial z}\right) = [\phi_1 : \cdots : \phi_m].$$

Note that the definition is independent of the choice of z. If M is minimal, then G is a holomorphic map. Moreover, since $\phi_1^2 + \cdots + \phi_m^2 = 0$, the image of G is contained in the quadric

$$Q_{m-1} = \{[w_1 : \cdots : w_m] \mid w_1^2 + \cdots + w_m^2 = 0\} \subset \mathbb{P}^{m-1}(\mathbb{C}).$$

The induced metric on M under the isothermal coordinate (u, v) is

$$ds^2 = (\mathbf{x}_u \cdot \mathbf{x}_u)(du^2 + dv^2) = 2\left(\left|\frac{\partial x_1}{\partial z}\right|^2 + \cdots + \left|\frac{\partial x_m}{\partial z}\right|^2\right)|dz|^2,$$

that is,

$$ds^2 = 2\|\tilde{G}\|^2|dz|^2,$$

where $\tilde{G}$ is any reduced representation of G. Its Gauss curvature by (3.6) is

$$K = -\frac{2}{\|\tilde{G}\|^2}\frac{\partial^2}{\partial z \partial \bar{z}} \log \|\tilde{G}\| = \frac{1}{\|\tilde{G}\|^6}\Big(\sum_{i<j} |g_i g_j' - g_j g_i'|^2\Big). \tag{5.3}$$

By definition a surface with a metric is said to be *flat* if and only if the Gauss curvature identically vanishes. By the above identity this means that $g_i/g_{i_0} \equiv$ constant for $1 \le i \le m$ with some i_0 such that $g_{i_0} \not\equiv 0$. Therefore G is a constant.

Theorem 5.1.6. *For a minimal surface M immersed in $\mathbb{R}^m$, M is flat, or, equivalently, the Gauss map of M is constant, if and only if it lies in a plane.*

Proof. The Gauss map of the surface that lies in a plane is obviously a constant. Conversely, we assume that G is constant. This means that every tangent plane $T_p(M)$ of M $(p \in M)$ is perpendicular to $m-2$ particularly linearly independent normal vectors $N_1, \dots, N_{m-2}$. We then have

$$\frac{\partial \mathbf{x}}{\partial u} \cdot N_k = \frac{\partial \mathbf{x}}{\partial v} \cdot N_k = 0$$

for $k = 1, \dots, m-2$ for all local coordinates u, v. Thus each value $\mathbf{x} \cdot N_k$ is a constant for $k = 1, \dots, m-2$, and so M lies in a plane. □

5.2 Value distributions of the (generalized) Gauss maps of complete minimal surfaces in $\mathbb{R}^m$

Let M be a minimal surface in $\mathbb{R}^m$, and let $\mathbf{x} = (x_1, \dots, x_m) : M \to \mathbb{R}^m$ be an immersion. Let $G : M \to \mathbb{P}^{m-1}(\mathbb{C})$ be a generalized Gauss map. The induced metric on M under the isothermal coordinate (u, v) is

$$ds^2 = (\mathbf{x}_u \cdot \mathbf{x}_u)(du^2 + dv^2) = 2\Big(\Big|\frac{\partial x_1}{\partial z}\Big|^2 + \cdots + \Big|\frac{\partial x_m}{\partial z}\Big|^2\Big)|dz|^2,$$

that is,

$$ds^2 = 2\|\tilde{G}\|^2|dz|^2,$$

where $\tilde{G}$ is any reduced representation of G. So the study of the value distributions of the (generalized) Gauss maps of complete minimal surfaces in $\mathbb{R}^m$ becomes the study of the holomorphic maps G from a complete Riemann surface M with respect to the metric $ds^2 = \|\tilde{G}\|^2|dz|^2$ into the complex projective space $\mathbb{P}^{m-1}(\mathbb{C})$.

Theorem 5.2.1 ([13]). *Let M be an open Riemann surface, and let $G : M \to \mathbb{P}^n(\mathbb{C})$ be a holomorphic map. Consider the conformal metric on M given by*

$$ds^2 = \|\tilde{G}\|^{2m}|\omega|^2,$$

where $\tilde{G}$ is a reduced representation of G, ω is a holomorphic 1-form, and $m \in \mathbb{Z}_{\geq 0}$. Assume that ds^2 is complete. If G is k-nondegenerate for some $1 \leq k \leq n$, then G can omit at most $mk(n - \frac{k-1}{2}) + 2n - k + 1$ hyperplanes in $\mathbb{P}^n(\mathbb{C})$ located in general position.

Note that the expression of the metric $ds^2 = \|\tilde{G}\|^{2m}|\omega|^2$ is independent of the choice of the representations in the following sense: if $\tilde{G}'$ and $\tilde{G}$ are two reduced representations of G, then $\tilde{G} = h\tilde{G}'$, where h is a nowhere vanishing holomorphic function. Hence $ds^2 = \|\tilde{G}\|^{2m}|\omega|^2 = \|\tilde{G}'\|^{2m}|\omega'|^2$, where $\omega' := h^m\omega$ is still a holomorphic 1-form.

Corollary 5.2.2. *Let M be an open Riemann surface, and let $G : M \to \mathbb{P}^n(\mathbb{C})$ be a holomorphic map. Consider the conformal metric on M given by*

$$ds^2 = \|\tilde{G}\|^{2m}|\omega|^2,$$

where $\tilde{G}$ is a reduced representation of G, ω is a holomorphic 1-form, and $m \in \mathbb{N}$. Assume that ds^2 is complete. Then:
(a) *When $m = 0$, if the image of G omits more than $2n$ hyperplanes in $\mathbb{P}^n(\mathbb{C})$ located in general position, then G is constant.*
(b) *When $m \geq 1$, if the image of G omits more than $\frac{n+1}{2}(mn + 2)$ hyperplanes in $\mathbb{P}^n(\mathbb{C})$ located in general position, then G is constant.*

To see how Theorem 5.2.1 implies Corollary 5.2.2, we note that when G is not constant, G must be k-nondegenerate for some $1 \leq k \leq n$. From Theorem 5.2.1 we know that G can omit at most $mk(n - \frac{k-1}{2}) + 2n - k + 1$ hyperplanes in $\mathbb{P}^n(\mathbb{C})$ located in general position. Noticing that when $m \geq 1$, $mk(n - \frac{k-1}{2}) + 2n - k + 1 \leq \frac{n+1}{2}(mn+2)$ for all $1 \leq k \leq n$, this gives a contradiction. When $m = 0$, the metric becomes $ds^2 = |\omega|^2$, which is flat and complete on M. In this case the universal cover of M is the whole complex plane $\mathbb{C}$ by Lemma 3.3.1. We thus may assume that G is a holomorphic map from $\mathbb{C}$ into $\mathbb{P}^n(\mathbb{C})$ by replacing G with $G \circ \Pi$, where $\Pi : \mathbb{C} \to M$ is the universal covering map. Therefore Corollary 5.2.2 is a generalization of Theorem 4.4.7.

Corollary 5.2.2 with $m = 1$ implies the following result we desired.

Theorem 5.2.3 ([67]). *The (generalized) Gauss map of a nonflat complete minimal surface immersed in a Euclidean n-space $\mathbb{R}^n$ can omit at most $n(n+1)/2$ hyperplanes in the complex projective $(n-1)$-space $\mathbb{P}^{n-1}(\mathbb{C})$ located in general position.*

Proof of Theorem 5.2.1. Assume that Theorem 5.2.1 does not hold, that is, the image of G omits hyperplanes $H_1, \ldots, H_q$ in $\mathbb{P}^n(\mathbb{C})$ in general position with $q > mk(n - \frac{k-1}{2}) + 2n - k + 1$. By taking the universal cover of M if necessary we can assume that M is simply connected. Then the uniformization theorem implies that M is conformally equivalent either to $\mathbb{C}$ or to the unit disc Δ. By Theorem 4.3.4 we know that a k-nondegenerate holomorphic map from $\mathbb{C}$ to $\mathbb{P}^n(\mathbb{C})$ cannot omit more than $2n - k + 1$ hyperplanes in general position. Hence, noticing that $mk(n - \frac{k-1}{2}) + 2n - k + 1 \geq 2n - k + 1$, G must be k-degenerate if M is conformally equivalent to $\mathbb{C}$, which contradicts with our assumption. So it remains to prove the theorem in the case that M is conformally equivalent to the unit disc Δ. We assume that M is the unit disc Δ.

Since G is k-nondegenerate, the image of G is contained in some k-dimensional projective subspace of $\mathbb{P}^n(\mathbb{C})$ but not in any subspace of dimension lower than k. Therefore by regarding $\mathbb{P}^k(\mathbb{C}) \subset \mathbb{P}^n(\mathbb{C})$ the map $G : \Delta \to \mathbb{P}^k(\mathbb{C})$ is linearly nondegenerate. Let $\tilde{H}_j := H_j \cap \mathbb{P}^k(\mathbb{C}), 1 \leq j \leq q$. Then these hyperplanes $\{\tilde{H}_j\}_{j=1}^q$ are in n-subgeneral position in $\mathbb{P}^k(\mathbb{C})$ since $\{H_j\}_{j=1}^q$ are, by assumption, in general position in $\mathbb{P}^n(\mathbb{C})$. We still denote $\tilde{H}_j$ as H_j. Hence it is reduced to the case that $G : \Delta \to \mathbb{P}^k(\mathbb{C})$ is linearly nondegenerate and the hyperplanes $H_j, 1 \leq j \leq q$, in $\mathbb{P}^k(\mathbb{C})$ are located in n-subgeneral position. Let $\tilde{G} = (g_0, g_1, \ldots, g_k)$ be a reduced representation of G, and let $\mathbf{a}_j := (a_{j0}, a_{j1}, \ldots, a_{jk})$ be the unit vectors associated with $H_j, 1 \leq j \leq q$. From the discussion in the last chapter, $\tilde{G}_k \not\equiv 0$, and none of the $\|\tilde{G}_s \llcorner \mathbf{a}_j\|, 0 \leq s \leq k, 1 \leq j \leq q$, vanishes identically, where, according to (4.58),

$$\|\tilde{G}_s \llcorner \mathbf{a}_j\|^2 = \sum_{0 \leq j_1 < \cdots < j_s \leq k} \left| \sum_{j_0 \neq j_1, \ldots, j_s} a_{jj_0} W(g_{j_0}, g_{j_1}, \ldots, g_{j_s}) \right|^2 .$$

Hence there exist $i_1, i_2, \ldots, i_s$ such that

$$\psi_{js} := \sum_{t \neq i_1, \ldots, i_s} a_{jt} W(g_t, g_{i_1}, \ldots, g_{i_s}) \tag{5.4}$$

does not vanish identically. Note that ψ_{js} is holomorphic and has only isolated zeros.

Let $\omega(j)$ be the Nochka weight associated with the hyperplane H_j for $1 \leq j \leq q$. By (ii) in Theorem 4.3.3

$$q - 2n + k - 1 = \theta\left(\sum_{j=1}^{q} \omega(j) - k - 1\right),$$

and

$$\theta \leq \frac{2n - k + 1}{k + 1}.$$

Hence

$$\frac{2(\sum_{j=1}^q \omega(j) - k - 1)}{mk(k+1)} = \frac{2(q - 2n + k - 1)}{\theta mk(k+1)} \geq \frac{2(q - 2n + k - 1)}{mk(2n - k + 1)}.$$

The condition $q > mk(n - \frac{k-1}{2}) + 2n - k + 1$ implies that

$$\frac{2(q - 2n + k - 1)}{mk(2n - k + 1)} > 1.$$

Therefore $\sum_{j=1}^q \omega(j) - k - 1 - \frac{mk}{2}(k+1) > 0$. Set

$$\chi := \sum_{j=1}^{q} \omega(j) - (k+1) - \frac{2q}{N}(k^2 + 2k - 1),$$

$$\lambda := \frac{m}{\chi}\left(\frac{1}{2}k(k+1) + \frac{2q}{N}\sum_{s=0}^{k} s^2\right).$$

Choose N such that the following inequality holds:

$$\frac{\sum_{j=1}^{q}\omega(j) - k - 1 - \frac{mk}{2}(k+1)}{\frac{2m}{q} + m\sum_{s=0}^{k} s^2 + k^2 + 2k - 1} < \frac{2q}{N} < \frac{\sum_{j=1}^{q}\omega(j) - k - 1 - \frac{mk}{2}(k+1)}{m\sum_{s=0}^{k} s^2 + k^2 + 2k - 1},$$

which implies that

$$0 < \lambda < 1, \qquad \frac{4m}{N\chi(1-\lambda)} > 1.$$

Write $\omega = hdz$, where h is a nowhere vanishing holomorphic function and define a new flat metric

$$d\sigma^2 = \left(\frac{\prod_{j=1}^{q}|a_{j0}g_0 + \cdots + a_{jk}g_k|^{\omega(j)}}{|W(g_0,\ldots,g_k)|^{1+\frac{2q}{N}}\prod_{j=1}^{q}(\prod_{s=0}^{k-1}|\psi_{js}|)^{\frac{4}{N}}}\right)^{\frac{2m}{(1-\lambda)\chi}} |h|^{\frac{2}{1-\lambda}}|dz|^2$$

on the set

$$M' := \Delta\backslash\left\{p \in \Delta \mid \text{either } W(g_0,\ldots,g_k)(p) = 0 \text{ or } \left(\prod_{j=1}^{q}\prod_{s=0}^{k-1}|\psi_{js}|\right)(p) = 0\right\}.$$

Here $W(g_0,\ldots,g_k)$ is the Wronskian determinant of $\tilde{G} = (g_0,\ldots,g_k)$.

Fix a point $p_0 \in M'$. By Lemma 3.3.1 there exists a local diffeomorphism Ψ of a disk $\Delta_R = \{w \in \mathbb{C} : |w| < R\}$ $(0 < R \le \infty)$ onto an open neighborhood of p_0 with $\Psi(0) = p_0$ such that Ψ is a local isometry. In addition, there exists a point a_0 with $|a_0| = 1$ such that the Ψ-image Γ_{a_0} of the curve $L_{a_0} = \{w = a_0 t : 0 < t < R\}$ is divergent in M'. Since G is not constant, $R < \infty$.

We claim that the Ψ-image Γ_{a_0} is indeed divergent to the boundary of Δ. To this end, we assume the contrary that the curve Γ_{a_0} is divergent to a point z_0 that satisfies either $\tilde{G}_k(z_0) = 0$ or $\psi_{js}(z_0) = 0$ for some s and j. Set

$$\Lambda = \left(\frac{\prod_{j=1}^{q}|a_{j0}g_0 + \cdots + a_{jk}g_k|^{\omega(j)}}{|W(g_0,\ldots,g_k)|^{1+\frac{2q}{N}}\prod_{j=1}^{q}(\prod_{s=0}^{k-1}|\psi_{js}|)^{\frac{4}{N}}}\right)^{\frac{m}{(1-\lambda)\chi}}.$$

Then, also noticing that z_0 is a pole of Λ with the multiplicities at least $\delta_0 = \frac{4m}{N\chi(1-\lambda_0)} (> 1)$ and using the fact that Ψ is an isometry,

$$R = \text{length of } L_{a_0} = \int_{\Gamma_{a_0}} \Lambda(z)|h(z)|^{\frac{1}{1-\lambda}}|dz| \geq c\int_{\Gamma_{a_0}} \frac{1}{|z-z_0|^{\delta_0}}|dz| = \infty.$$

This gives a contradiction. Therefore $\Gamma_{a_0} = \Psi(L_{a_0})$ is divergent to the boundary of Δ.

Our goal now is to show that Γ_{a_0} has finite length in the (original) metric $\|\tilde{G}\|^{2m}|\omega|^2$, contradicting the completeness of the (original) metric $\|\tilde{G}\|^{2m}|\omega|^2$. To do so, we use the isometry property of Ψ, that is, if we let $z = \Psi(w)$, then

$$|dw| = d\sigma = \left(\frac{\prod_{j=1}^q |a_{j0}g_0 + \cdots + a_{jk}g_k|^{\omega(j)}}{|W(g_0,\ldots,g_k)|^{1+\frac{2q}{N}}\prod_{j=1}^q(\prod_{s=0}^{k-1}|\psi_{js}|)^{\frac{4}{N}}}\right)^{\frac{m}{(1-\lambda)\chi}}|h|^{\frac{1}{1-\lambda}}|dz|. \tag{5.5}$$

To change the coordinate z into w in the right-hand side of (5.5), we introduce the following functions defined on $\{w : |w| < R\}$: $f_s(w) := g_s(\Psi(w))$, $0 \leq s \leq k$, and $f := [f_0 : f_1 : \cdots : f_k]$, that is, $f(w) = G(\Psi(w))$. For $1 \leq j \leq q$ and $0 \leq s \leq k$, we let

$$\varphi_{js} := \sum_{t\neq i_1,\ldots,i_s} a_{jt}W(f_t, f_{i_1},\ldots,f_{i_s}),$$

where $(i_1,\ldots,i_s)$ is the index in the definition of ψ_{js} in (5.4). Denote by $\mathbf{F}_s(w)$ and $\tilde{G}_s(z)$ the sth associated maps of f and G, respectively. Then it is obvious from the definition that

$$\mathbf{F}_s(w) = \tilde{G}_s(z)\left(\frac{dz}{dw}\right)^{s(s+1)/2}, \quad 0 \leq s \leq k.$$

In particular, noticing that $|W(f_0,\ldots,f_k)| = \|\mathbf{F}_k\|$, we have

$$|W(f_0,\ldots,f_k)| = |W(g_0,\ldots,g_k)|\left|\frac{dz}{dw}\right|^{k(k+1)/2}.$$

Thus, if we let $\mathbf{F} := (f_0,\ldots,f_k)$, then (5.5) becomes

$$\begin{aligned}|dw| &= \left(\frac{\prod_{j=1}^q |a_{j0}f_0 + \cdots + a_{jk}f_k|^{\omega(j)}}{|W(f_0,\ldots,f_k)|^{1+\frac{2q}{N}}\prod_{j=1}^q(\prod_{s=0}^{k-1}|\varphi_{js}|)^{\frac{4}{N}}}\right)^{\frac{m}{(1-\lambda)\chi}} \\ &\quad\times\left|\frac{dz}{dw}\right|^{\frac{m((1+\frac{2q}{N})\frac{k(k+1)}{2}+\frac{2q}{N}\sum_{s=0}^{k-1}s(s+1))}{(1-\lambda)\chi}}|h|^{\frac{1}{1-\lambda}}|dz| \\ &= \left(\frac{\prod_{j=1}^q |a_{j0}f_0 + \cdots + a_{jk}f_k|^{\omega(j)}}{|W(f_0,\ldots,f_k)|^{1+\frac{2q}{N}}\prod_{j=1}^q(\prod_{s=0}^{k-1}|\varphi_{js}|)^{\frac{4}{N}}}\right)^{\frac{m}{(1-\lambda)\chi}}\left|\frac{dz}{dw}\right|^{\frac{\lambda}{1-\lambda}}|h|^{\frac{1}{1-\lambda}}|dz|,\end{aligned}$$

that is,

$$\frac{|dw|}{|dz|} = \left(\frac{\prod_{j=1}^{q} |a_{j0}f_0 + \cdots + a_{jk}f_k|^{\omega(j)}}{|W(f_0, \ldots, f_k)|^{1+\frac{2q}{N}} \prod_{j=1}^{q} (\prod_{s=0}^{k-1} |\varphi_{js}|)^{\frac{4}{N}}} \right)^{\frac{m}{\chi}} |h|.$$

Denote by $l(\Gamma_{a_0})$ the length of the curve Γ_{a_0} with respect to the metric $\|\tilde{G}\|^{2m}|\omega|^2$. Then using $|\varphi_{js}| \le |\mathbf{F}_s(H_j)|$ and noticing that $\|\mathbf{F}_k\|^2 = |W(f_0, \ldots, f_k)|$ and $\mathbf{F}(H_j) = a_{j0}f_0 + \cdots + a_{jk}f_k$, we get

$$\begin{aligned} l(\Gamma_{a_0}) &= \int_{\Gamma_{a_0}} \|\tilde{G}\|^m |\omega| = \int_{L_{a_0}} \Psi^*(\|\tilde{G}\|^m |\omega|) = \int_{L_{a_0}} \|\mathbf{F}\|^m |h \circ \Psi| \frac{|dz|}{|dw|} |dw| \\ &\le \int_{L_{a_0}} \left(\frac{\|\mathbf{F}\|^{\chi} \|\mathbf{F}_k\|^{1+\frac{2q}{N}} \prod_{j=1}^{q} (\prod_{s=0}^{k-1} |\mathbf{F}_s(H_j)|)^{\frac{4}{N}}}{\prod_{j=1}^{q} |\mathbf{F}(H_j)|^{\omega(j)}} \right)^{\frac{m}{\chi}} |dw|. \end{aligned}$$

Applying Lemma 4.4.9, we get

$$l(\Gamma_{a_0}) \le C \int_0^R \left(\frac{2R}{R^2 - |w|^2} \right)^{\lambda} |dw| < \infty$$

since $0 < \lambda < 1$. This contradicts the completeness assumption of the metric $\|\tilde{G}\|^{2m}|\omega|^2$, and so we complete the proof of the theorem. □

5.3 The Gauss curvature estimate

In this subsection, we study the minimal surfaces without the completeness condition. In particular, we prove Theorem 5.3.2, which generalizes Theorem 5.2.1. To do so, we first prove a weaker result.

Theorem 5.3.1 ([13]). *Let M be an open Riemann surface, and let $G : M \to \mathbb{P}^n(\mathbb{C})$ be a nonconstant holomorphic map. Consider the conformal metric on M given by*

$$ds^2 = \|\tilde{G}\|^{2m}|\omega|^2,$$

where $\tilde{G}$ is a reduced representation of G, ω is a holomorphic 1-form on M, and $m \in \mathbb{N}$. Assume that G omits a neighborhood U of a hyperplane in $\mathbb{P}^n(\mathbb{C})$. Then there exists a constant C, which only depends on U (and not on G and M), such that

$$|K(p)|^{\frac{1}{2}} d(p) \le C, \tag{5.6}$$

where $K(p)$ is the Gauss curvature of the surface at p, and $d(p)$ is the geodesic distance from p to the boundary of M.

If we assume that M is complete with respect to the induced metric, then $d(p) = \infty$ for any $p \in M$. Hence $K \equiv 0$, so the minimal must lie on a plane.

We derive an expression for the Gauss curvature K of M with the conformal metric $ds^2 = \|\tilde{G}\|^{2m}|\omega|^2$. Since ω is a holomorphic 1-form, from (3.6) we have that the Gauss curvature of ds^2 is

$$K = -m\frac{\Delta \log \|\tilde{G}\|}{\|\tilde{G}\|^{2m}|h|^2} = -2m\frac{\sum_{0\le i<j\le n}|g_i g_j' - g_i' g_j|^2}{\|\tilde{G}\|^{2m+4}|h|^2}, \tag{5.7}$$

where $\omega = hdz$ and $\tilde{G} = (g_0, \dots, g_n)$. Define the meromorphic functions

$$\psi_j(z) = \frac{g_j(z)}{g_n(z)}, \quad j = 0, 1, \dots, n-1. \tag{5.8}$$

Then we have

$$\psi_i\psi_j' - \psi_i'\psi_j = \frac{g_i g_j' - g_i' g_j}{g_n^2}, \quad i, j = 0, 1, \dots, n-1.$$

Hence (5.7) can be rewritten as

$$K = -2m\frac{\sum_{0\le i<j\le n-1}|\psi_i\psi_j' - \psi_i'\psi_j|^2 + \sum_{j=0}^{n-1}|\psi_j'|^2}{|g_n|^{2m}(1 + \sum_{j=0}^{n-1}|\psi_j|^2)^{m+2}|h|^2}. \tag{5.9}$$

Proof of Theorem 5.3.1. By taking the universal cover of M we may assume that M is simply connected. So M is conformally equivalent either to $\mathbb{C}$ or to the unit disc Δ. If M is conformally equivalent to $\mathbb{C}$, then by Corollary 4.3.5 G must be constant, which contradicts our assumption. So we may assume that $M = \Delta$. For any (fixed) $p \in M$, we assume that p corresponds to the point $z = 0$. Without loss of generality, we assume that G omits a neighborhood of the hyperplane $H = \{y_n = 0\}$ in $\mathbb{P}^n(\mathbb{C})$, and thus there is a positive constant $1 > \epsilon > 0$ such that

$$\frac{|g_n|}{\|\tilde{G}\|} \ge \epsilon \tag{5.10}$$

on Δ, where $\tilde{G} = (g_0, \dots, g_n)$. Let $\psi_j := \frac{g_j}{g_n}$. Then this implies that

$$\sum_{j=0}^{n-1}|\psi_j(z)|^2 \le \frac{1}{\epsilon^2} - 1. \tag{5.11}$$

Denote

$$C_j := |\psi_j(0)|, \quad D_j := |\psi_j'(0)|, \quad M_j := \sup_{|z|<1}|\psi_j(z)|.$$

Applying the Schwarz–Pick lemma to the function ψ_j/M_j, we get

$$D_j \le M_j\left(1 - \frac{C_j^2}{M_j^2}\right) = M_j\eta_j, \quad \text{where } \eta_j := 1 - \frac{C_j^2}{M_j^2} < 1.$$

Furthermore, it follows from (5.11) that for $0 \le j \le n-1$,

$$C_j^2 \le M_j^2 \le \frac{1}{\epsilon^2} - 1, \quad D_j^2 \le \left(\frac{1}{\epsilon^2} - 1\right)\eta_j^2. \tag{5.12}$$

By the Cauchy–Schwarz inequality we have

$$\begin{aligned}\sum_{0\le i<j\le n-1} |\psi_i(0)\psi_j'(0) - \psi_i'(0)\psi_j(0)|^2 &\le \sum_{0\le i<j\le n-1} (C_iD_j + C_jD_i)^2 \\ &\le \sum_{0\le i<j\le n-1} 2(C_i^2D_j^2 + C_j^2D_i^2) \le 2\sum_{i=0}^{n-1} C_i^2 \sum_{j=0}^{n-1} D_j^2.\end{aligned}$$

This, together with (5.12), yields

$$\sum_{0\le i<j\le n-1} |\psi_i(0)\psi_j'(0) - \psi_i'(0)\psi_j(0)|^2 + \sum_{j=0}^{n-1}|\psi_j'(0)|^2 \le 2n\left(\frac{1}{\epsilon^2} - 1\right)\left(1 + \sum_{i=0}^{n-1} C_i^2\right).$$

Thus, using (5.9), we get

$$|K(0)| \le \frac{4mn}{|g_n^{2m}(0)h^2(0)|}\left(\frac{1}{\epsilon^2} - 1\right). \tag{5.13}$$

Next, we estimate $d(0)$, which is the geodesic distance from 0 to the boundary of Δ. Let γ be a simple curve from 0 to the boundary of Δ. With the help of (5.10) the length of the curve γ on M satisfies

$$l(\gamma) = \int_\gamma ds = \int_\gamma \|\tilde{G}\|^m |h||dz| \le \frac{1}{\epsilon^m}\int_\gamma |g_n^m||h||dz|.$$

Since the metric $d\sigma_0 := |g_n^m||h||dz|$ is flat, by Lemma 3.3.1 there exists a local diffeomorphism Ψ of a disk $\Delta_R = \{w \in \mathbb{C} : |w| < R\}$ $(0 < R \le \infty)$ onto an open neighborhood of 0 with $\Psi(0) = 0$ such that Ψ is a local isometry. Furthermore, there exists w_0 with $|w_0| = R$ such that if Γ is the image of line $L := \{w = w_0 t, 0 < t < 1\}$ under Ψ, then the curve Γ is divergent in Δ. From Corollary 4.3.5 we know that $R < \infty$ since G is nonconstant. Regarding Ψ as a holomorphic map from Δ_R into the unit disc Δ, by Schwarz's lemma, we have $|\Psi'(0)| \le 1/R$. Thus

$$d(0) = \inf_\gamma \int_\gamma \|\tilde{G}\|^m |h||dz| \le \int_\Gamma \|\tilde{G}\|^m |h||dz| \le \frac{1}{\epsilon^m}\int_\Gamma |g_n^m||h||dz|$$

$$= \frac{1}{\epsilon^m}\int_L |dw| = \frac{R}{\epsilon^m} \le \frac{1}{\epsilon^m|\Psi'(0)|} = \frac{|g_n^m(0)h(0)|}{\epsilon^m}.$$

By combing this inequality with (5.13) we get

$$|K(0)|d^2(0) \le 4mn\frac{1-\epsilon^2}{\epsilon^{2m+2}}.$$

This completes the proof. □

Theorem 5.3.2. *Let M be an open Riemann surface, and let $G : M \to \mathbb{P}^n(\mathbb{C})$ be a nonconstant holomorphic map. Consider the conformal metric on M given by*

$$ds^2 = \|\tilde{G}\|^{2m}|\omega|^2,$$

where $\tilde{G}$ is a reduced representation of G, ω is a holomorphic 1-form, and $m \in \mathbb{N}$. Assume that G omits more than $\frac{n+1}{2}(mn+2)$ hyperplanes in $\mathbb{P}^n(\mathbb{C})$ located in general position. Then there exists a constant C depending only on the set of omitted hyperplanes such that

$$|K(p)|^{\frac{1}{2}}d(p) \le C, \tag{5.14}$$

where $K(p)$ is the Gauss curvature of M at p with respect to ds^2, and $d(p)$ is the geodesic distance from p to the boundary of M.

To prove Theorem 5.3.2, we need some preparational results.

Lemma 5.3.3. *Let M be a Riemann surface. Let $G^{(l)} : M \to \mathbb{P}^n(\mathbb{C})$ be a sequence of holomorphic maps converging uniformly on every compact subset of M to a holomorphic map $G : M \to \mathbb{P}^n(\mathbb{C})$. Let H_1, H_2 be two distinct hyperplanes in $\mathbb{P}^n(\mathbb{C})$. Let $p \in M$ be such that for all $z \in U_p$, $\tilde{G}(H_2)(z) \ne 0$ and $\tilde{G^{(l)}}(H_2)(z) \ne 0$ for l large enough, where U_p is a neighborhood of p, $\tilde{G}$ is a reduced representation of G, and $\tilde{G}^{(l)}$ is a reduced representative of $G^{(l)}$ on U_p. Then $\{\tilde{G}^{(l)}(H_1)/\tilde{G}^{(l)}(H_2)\}$ converges uniformly on U_p to $\frac{\tilde{G}(H_1)}{\tilde{G}(H_2)}$.*

Proof. Let $H_j := \{[w_0 : \cdots : w_n] \mid a_{j0}w_0 + \cdots + a_{jn}w_n = 0\}$ for $j = 1, 2$. Then $\phi := \frac{a_{10}w_0+\cdots+a_{1n}w_n}{a_{20}w_0+\cdots+a_{2n}w_n}$ is a rational function on $\mathbb{P}^n(\mathbb{C})$. Since $G^{(l)}$ converges uniformly on U_p to G and for all $z \in U_p$, $\tilde{G}(H_2)(z) \ne 0$ and $\tilde{G^{(l)}}(H_2)(z) \ne 0$ for l large enough, we have that $\{\tilde{G}^{(l)}(H_1)/\tilde{G}^{(l)}(H_2)\}$ converges uniformly on U_p to $\frac{\tilde{G}(H_1)}{\tilde{G}(H_2)}$. □

Lemma 5.3.4. *Let ds_n^2 be a sequence of conformal metrics on the unit disk D with curvatures satisfying $-1 \le K_n \le 0$. Suppose that D is a geodesic disk of radius R_n with respect to the metric ds_n^2, where $R_n \to \infty$, and that the metrics ds_n^2 converge, uniformly on compact sets, to a metric ds^2. Then all distances to the origin with respect to ds^2 are greater than or equal to the corresponding hyperbolic distances in D. In particular, ds^2 is complete.*

Proof. For any point z in D, let $\rho_n(z)$ be the distance from 0 to z in the metric ds_n^2, and let $\rho(z)$ be the distance in the limit metric ds^2. Let $|z| = r_n$ be the circle in D of hyperbolic radius R_n. Explicitly,

$$R_n = \log \frac{1 + r_n}{1 - r_n}.$$

Making the change of parameter $w = r_n z$, we may apply Lemma 2.3.11 to the induced metric in $|w| < r_n$ and conclude that

$$\rho_n(z) \geq \log \frac{1 + |w|}{1 - |w|} = \log \frac{1 + r_n|z|}{1 - r_n|z|}.$$

As $n \to \infty$, we have $R_n \to \infty$ and $r_n \to 1$. Hence by the uniform convergence on compact sets we have

$$\rho(z) = \lim_{n\to\infty} \rho_n(z) \geq \lim_{r_n \to 1} \log \frac{1 + r_n|z|}{1 - r_n|z|} = \log \frac{1 + |z|}{1 - |z|},$$

which proves the lemma. □

The following proposition plays a central role in the proof of Theorem 5.3.2.

Proposition 5.3.5. *Let M be an open simply connected Riemann surface, and let $G^{(l)} : M \to \mathbb{P}^n(\mathbb{C})$ be a sequence of holomorphic maps. Fix a global reduced representation $\tilde{G}^{(l)}$ of $G^{(l)}$. Let*

$$ds_l^2 = \|\tilde{G}^{(l)}\|^{2m} |dz|^2,$$

where $m \in \mathbb{N}$. Denote by K_l the Gauss curvature of M with respect to ds_l^2. Assume that $\{G^{(l)}\}$ converges to a nonconstant holomorphic map G uniformly on every compact subset of M and $\{|K_l|\}$ is uniformly bounded. Then one of the following statements must be true.
(i) *There is a subsequence $\{K_{l_i}\}$ of $\{K_l\}$ that converges to zero;*
(ii) *For each $0 \leq j \leq n$, there exists a subsequence $\{g_j^{(l_i)}\}$ of $\{g_j^{(l)}\}$ that converges to a holomorphic function h_j on M. Furthermore, $h_0, \dots, h_n$ have no common zeros.*

We remark that Proposition 5.3.5 also holds for the metric $ds_l^2 = \|\tilde{G}^{(l)}\|^{2m} |\omega^{(l)}|^2$, where $\omega^{(l)} = h^{(l)} dz$ is a holomorphic 1-form. Indeed, we can just consider a reduced representation $\tilde{G}^{(l)} = (g_0^{(l)}/h^{(l)}, g_1^{(l)}/h^{(l)}, \dots, g_n^{(l)}/h^{(l)})$ of $G^{(l)}$, instead of $(g_0^{(l)}, g_1^{(l)}, \dots, g_n^{(l)})$.

Proof of Proposition 5.3.5. Recall that, by (5.7), for any $G : M \to \mathbb{P}^n(\mathbb{C})$ with the reduced representation $\tilde{G}$, the metric $ds^2 := \|\tilde{G}\|^{2m}|dz|^2$ on M has the Gauss curvature expression

$$K = -m \frac{\Delta \log \|\tilde{G}\|}{\|\tilde{G}\|^{2m}} = -2m \frac{\|\tilde{G} \wedge \tilde{G}'\|^2}{\|\tilde{G}\|^{2m+4}} = -2m \frac{\sum_{0 \leq i < j \leq n} |g_i g_j' - g_i' g_j|^2}{\|\tilde{G}\|^{2m+4}}. \tag{5.15}$$

Without loss of generality, we may assume that $|K_l| \leq 1$ for all l. By the assumption, $\{G^{(l)}\}$ converges uniformly on every compact subset of M to the nonconstant holomorphic map $G : M \to \mathbb{P}^n(\mathbb{C})$. Let $\tilde{G} = (g_0, \ldots, g_n)$ be a reduced representation of G. Since G is not constant, $G(M)$ is not contained in some coordinate hyperplane. Without loss of generality, we assume that $G(M) \not\subset H_0$, where $H_0 := \{[y_0 : \cdots : y_n] \in \mathbb{P}^n(\mathbb{C}) \mid y_0 = 0\}$.

Set

$$M_0 = \{p \in M \mid G(p) \notin H_0, \tilde{G}(p) \wedge \tilde{G}'(p) \neq 0\}.$$

Obviously, $M \setminus M_0$ is a discrete set that consists of zeros of g_0 and the common zeros of $g_j g_k' - g_k g_j', 0 \leq j, k \leq n$.

Let $p \in M_0$. Then there exists a neighborhood U_p of p such that $g_0 \neq 0$ on U_p and

$$2c \leq \frac{2m\|\tilde{G} \wedge \tilde{G}'\|^2/|g_0|^4}{(1 + \sum_{j=1}^n |\frac{g_j}{g_0}|^2)^{m+2}} = \frac{2m \sum_{j<k} |\frac{g_j}{g_0}(\frac{g_k}{g_0})' - \frac{g_k}{g_0}(\frac{g_j}{g_0})'|^2}{(1 + \sum_{j=1}^n |\frac{g_j}{g_0}|^2)^{m+2}}$$

on U_p for some constant $c > 0$. Because $G^{(l)} \to G$ uniformly on U_p, we have $G^{(l)}(z) \notin H_0$ for l large enough, and thus for $1 \leq j \leq n$, $\{g_j^{(l)}/g_0^{(l)}\}$ converges uniformly to g_j/g_0 on U_p. Hence the above inequality implies that for l large enough,

$$\frac{2m \sum_{j<k} |\frac{g_j^{(l)}}{g_0^{(l)}}(\frac{g_k^{(l)}}{g_0^{(l)}})' - \frac{g_k^{(l)}}{g_0^{(l)}}(\frac{g_j^{(l)}}{g_0^{(l)}})'|^2}{(1 + \sum_{j=1}^n |\frac{g_j^{(l)}}{g_0^{(l)}}|^2)^{m+2}} \geq c. \tag{5.16}$$

On the other hand, by (5.15)

$$|K_l| = \frac{2m \sum_{j<k} |\frac{g_j^{(l)}}{g_0^{(l)}}(\frac{g_k^{(l)}}{g_0^{(l)}})' - \frac{g_k^{(l)}}{g_0^{(l)}}(\frac{g_j^{(l)}}{g_0^{(l)}})'|^2}{|g_0^{(l)}|^{2m}(1 + \sum_{j=1}^n |\frac{g_j^{(l)}}{g_0^{(l)}}|^2)^{m+2}} \leq 1. \tag{5.17}$$

Combination of (5.16) and (5.17) yields $|g_0^{(l)}|^{2m} \geq c$ on U_p for large enough l. Hence, by Montel's theorem, $\{g_0^{(l)}\}$ forms a normal family on U_p. Since $M - M_0$ is discrete, by taking a subsequence if necessary we can assume that $\{g_0^{(l)}\}$ converges to a holomorphic function h_0 on M_0 or to infinity uniformly on every subset of M_0.

We first claim that the case that $\{g_0^{(l)}\}$ converges to infinity uniformly on every subset of M_0 leads to conclusion (i) in the proposition. Indeed, assuming that $\{g_0^{(l)}\} \to \infty$ on M_0, for the points $p \in M_0$, by (5.17) we have

$$|K_l(p)| = \frac{2m \sum_{j<k} |\frac{g_j^{(l)}}{g_0^{(l)}}(\frac{g_k^{(l)}}{g_0^{(l)}})' - \frac{g_k^{(l)}}{g_0^{(l)}}(\frac{g_j^{(l)}}{g_0^{(l)}})'|^2}{|g_0^{(l)}|^{2m}(1 + \sum_{j=1}^n |\frac{g_j^{(l)}}{g_0^{(l)}}|^2)^{m+2}} \to 0.$$

For the points $p \notin M_0$ with $G(p) \notin H_0$, in a small disc $\Delta_{2\epsilon}(p)$ centered at p with radius 2ϵ, $G^{(l)}(z) \notin H_0$ for large enough l and all $z \in \Delta_{2\epsilon}(p)$, that is, $g_0^{(l)}$ is a nowhere vanishing holomorphic function on $\Delta_{2\epsilon}(p)$, and $\{g_0^{(l)}\}$ converges to infinity uniformly on $\partial\Delta_\epsilon(p)$. By the maximum principle we conclude that $\{g_0^{(l)}\}$ converges to infinity on $\Delta_\epsilon(p)$. Hence $|K_l(p)| \to 0$. Finally, for p with $G(p) \in H_0$, since $G(p)$ is not contained in some coordinate hyperplane, we assume that $G(p) \notin H_1$, where $H_1 := \{[y_0 : \cdots : y_n] \in \mathbb{P}^n(\mathbb{C}) \mid y_1 = 0\}$. Therefore on a small disc $\Delta_{2\epsilon}(p)$ centered at p, $g_1(z) \neq 0$, $g_1^{(l)}(z) \neq 0$ for all $z \in \Delta_{2\epsilon}(p)$, and $g_0^{(l)}$ and g_0 have no zeros on the neighborhood of $\partial\Delta_\epsilon(p)$ for l large enough. Hence it follows from Lemma 5.3.3 that $\{g_1^{(l)}/g_0^{(l)}\}$ converges uniformly to $\frac{g_1}{g_0}$ on $\partial\Delta_\epsilon(p)$. Since $\{g_0^{(l)}\}$ converges to infinity uniformly on every subset of M_0, this implies that the sequence $\{\frac{g_1^{(l)}}{g_0^{(l)}} g_0^{(l)}\}$, or, equivalently, $\{g_1^{(l)}\}$, converges uniformly to infinity on $\partial\Delta_\epsilon(p) \subset \Delta_{2\epsilon}(p)$. Again by the maximum principle, we conclude that $\{g_1^{(l)}\}$ converges to infinity on $\Delta_\epsilon(p)$. Hence

$$|K_l(p)| = \frac{2m \sum_{j<k} |\frac{g_j^{(l)}(p)}{g_1^{(l_i)}(p)} (\frac{g_k^{(l)}}{g_1^{(l)}})'(p) - \frac{g_k^{(l)}(p)}{g_1^{(l)}(p)} (\frac{g_j^{(l)}}{g_1^{(l)}})'(p)|^2}{|g_1^{(l)}(p)|^{2m} (1 + \sum_{j \neq 1} |\frac{g_j^{(l)}(p)}{g_1^{(l)}(p)}|^2)^{m+2}} \to 0.$$

In conclusion, we proved that $|K_l(p)| \to 0$ for all $p \in M$ if $\{g_0^{(l)}\}$ converges to infinity uniformly on every compact subset of M_0.

It remains to consider the case where $\{g_0^{(l)}\}$ converges to a holomorphic function h_0 uniformly on every compact subset of M_0. First, we show that h_0 can be extended to a holomorphic function on M. Indeed, let $p \in M \setminus M_0$. Then there exists a small disc $\Delta_{2\epsilon}(p)$ centered at p such that $\{g_0^{(l)}\}$ converges to h_0 uniformly on $\partial\Delta_\epsilon(p)$. Since $g_0^{(l)}$ is holomorphic, by using the maximum principle we see that $\{g_0^{(l)}\}$ is uniformly bounded on $\Delta_\epsilon(p)$. Namely, $\{g_0^{(l)}\}$ is normal on $\Delta_\epsilon(p)$. So there exists a subsequence of $\{g_0^{(l)}\}$, say itself, converging uniformly on $\Delta_\epsilon(p)$. Therefore h_0 can be extended to a holomorphic function on M, and $\{g_0^{(l)}\}$ converges to h_0 on M.

We now prove that $\{g_j^{(l)}\}$ converges uniformly to holomorphic functions h_j on M for $1 \leq j \leq n$. For $p \in M$ such that $G(p) \notin H_0$, there exists a neighborhood U_p of p such that g_0 and $g_0^{(l)}$ have no zeros on U_p for l large enough. Therefore, by Lemma 5.3.3, $\{g_j^{(l)}/g_0^{(l)}\}$ converges uniformly on U_p. Noticing that $g_j^{(l)} = \frac{g_j^{(l)}}{g_0^{(l)}} g_0^{(l)}$ and $\{g_0^{(l)}\}$ converges uniformly to h_0 on U_p, we know that $\{g_j^{(l)}\}$ converges to some holomorphic function h_j uniformly on U_p for $1 \leq j \leq n$. For the points p with $G(p) \in H_0$, take $\Delta_{2\epsilon}(p)$ centered at p such that g_0 and $g_0^{(l)}$ have no zeros on the neighborhood of $\partial\Delta_\epsilon(p)$ for l large enough. Then, as we just proved, $\{g_j^{(l)}\}$ is uniformly convergent on $\partial\Delta_\epsilon(p)$ for $1 \leq j \leq n$. Since $g_j^{(l)}$ is holomorphic, by the maximum principle, $\{g_j^{(l)}\}$ is uniformly bounded on $\Delta_\epsilon(p)$, and thus $\{g_j^{(l)}\}$ is normal on $\Delta_\epsilon(p)$. So there exists a subsequence of $\{g_j^{(l)}\}$, say itself, which is convergent on $\Delta_\epsilon(p)$. Let h_j be its limit function. Then h_j is holomorphic on $\Delta_\epsilon(p)$.

Finally, we show that h_j, $0 \le j \le n$, have no common zeros. Since G is nonconstant, for any given $p \in M$, there exist s with $0 \le s \le n$ and a small neighborhood U_p of p such that g_s and $g_s^{(l)}$ have no zeros on U_p for l large enough. Since $\{g_s^{(l)}\}$ converges uniformly h_s on U_p, it follows from Theorem 2.1.8 (Hurwitz's theorem) that either h_s has no zeros or $h_s \equiv 0$ on U_p. If h_s has no zeros, then we are done. Assuming that $h_s \equiv 0$ on U_p, that is, $\{g_s^{(l)}\}$ converges to zero on U_p, pick a point $q \in U_p$ such that $q \in M_0$. Since $g_s(z) \ne 0$ and $g_s^{(l)}(z) \ne 0$ on U_p for l large enough, we have

$$|K_l(q)| = \frac{2m \sum_{j<k} |\frac{g_j^{(l)}(q)}{g_s^{(l)}(q)}(\frac{g_k^{(l)}}{g_s^{(l)}})'(q) - \frac{g_k^{(l)}(q)}{g_s^{(l)}(q)}(\frac{g_j^{(l)}}{g_s^{(l)}})'(q)|^2}{|g_s^{(l)}(q)|^{2m}(1 + \sum_{j\ne s} |\frac{g_j^{(l)}(q)}{g_s^{(l)}(q)}|^2)^{m+2}}$$

$$= \left|\frac{g_s(q)}{g_s^{(l)}(q)}\right|^{2m} \cdot \frac{2m \sum_{j<k} |\frac{g_j^{(l)}(q)}{g_s^{(l)}(q)}(\frac{g_k^{(l)}}{g_s^{(l)}})'(q) - \frac{g_k^{(l)}(q)}{g_s^{(l)}(q)}(\frac{g_j^{(l)}}{g_s^{(l)}})'(q)|^2}{|g_s(q)|^{2m}(1 + \sum_{j\ne s} |\frac{g_j^{(l)}(q)}{g_s^{(l)}(q)}|^2)^{m+2}}.$$

Notice that

$$\frac{2m \sum_{j<k} |\frac{g_j^{(l)}(q)}{g_s^{(l)}(q)}(\frac{g_k^{(l)}}{g_s^{(l)}})'(q) - \frac{g_k^{(l)}(q)}{g_s^{(l)}(q)}(\frac{g_j^{(l)}}{g_s^{(l)}})'(q)|^2}{|g_s(q)|^{2m}(1 + \sum_{j\ne s} |\frac{g_j^{(l)}(q)}{g_s^{(l)}(q)}|^2)^{m+2}} \to \frac{2m\|\tilde{G}(q) \wedge \tilde{G}'(q)\|^2}{\|\tilde{G}(q)\|^{2(m+2)}}$$

as $l \to \infty$. Thus $|K_l(q)| \to \infty$ since $g_s^{(l)}(q) \to 0$. This contradicts with the assumption that $\{|K_l|\}$ is uniformly bounded. Hence $h_0, h_1, \ldots, h_n$ have no common zeros. This completes the proof of the proposition. □

Proof of Theorem 5.3.2. If ds^2 is complete, then by Corollary 5.2.2, G is a constant map, which contradicts our assumption. So we just consider the case that ds^2 is not complete on M.

Assume that Theorem 5.3.2 does not hold. Then there exist a sequence of holomorphic maps $G^{(l)} : M_l \to \mathbb{P}^n(\mathbb{C})$ and a sequence of points $p_l \in M_l$ such that $|K_l(p_l)|d_l^2(p_l) \to \infty$ and $G^{(l)}$ omits a fixed set of more than $\frac{n+1}{2}(mn + 2)$ hyperplanes in general position, where M_l are open Riemann surfaces.

We claim that the surfaces M_l can be chosen such that

$$K_l(p_l) = -\frac{1}{4} \quad \text{and} \quad -1 \le K_l \le 0 \tag{5.18}$$

on M_l for all l. Indeed, without loss of generality, assume that M_l is a geodesic disk centered at p_l. Let

$$M_l^* = \{p \in M_l : d_l(p, p_l) \le d_l(p_l)/2\}.$$

Then $\{|K_l|\}$ is uniformly bounded on M_l^*, and $d_l^*(p)$ tends to zero as p tends to ∂M_l^*, where $d_l^*(p)$ is the distance from p to ∂M_l^*. Hence there exists an interior point p_l^* of M_l^* such that

$$|K_l(p_l^*)|(d_l^*(p_l^*))^2 = \max_{p \in M_l^*} |K_l(p)|(d_l^*(p))^2,$$

and thus

$$|K_l(p_l^*)|(d_l^*(p_l^*))^2 \geq |K_l(p_l)|(d_l^*(p_l))^2 = \frac{1}{4}|K_l(p_l)|(d_l(p_l))^2 \to \infty.$$

So we can replace M_l by M_l^* with $|K_l(p_l^*)|d_l^{*2}(p_l^*) \to \infty$. We rescale M_l^* to make $K_l(p_l^*) = -\frac{1}{4}$. Due to the invariance under scaling of the quantity $|K(p)|d^2(p)$, we have $d_l^*(p_l^*) \to \infty$. Here, without causing confusion, we use the same notation d_l^* to denote the geodesic distance with respect to the rescaled metric. Again, we can assume that M_l^* geodesic disk centered at p_l^* and let

$$M_l^{**} = \{p \in M_l^* : d_l(p, p_l^*) \leq d_l^*(p_l^*)/2\}.$$

Then $p \in M_l^{**}$ implies that $d_l^*(p) \geq \frac{d_l^*(p_l^*)}{2}$ and

$$|K_l(p)|(d_l^*(p))^2 \leq |K_l(p_l^*)|(d_l^*(p_l^*))^2 = \frac{1}{4}(d_l^*(p_l^*))^2 \leq (d_l^*(p))^2,$$

that is, $|K_l(p)| \leq 1$ for $p \in M_l^{**}$. Furthermore,

$$d_l^{**}(p_l^*) = d_l(p_l^*, \partial M_l^{**}) = \frac{d_l^*(p_l^*)}{2} \to \infty. \tag{5.19}$$

This proves the claim.

Note that the Gauss curvature does not change if we go to the universal cover. So by taking universal cover of M_l if necessary we can assume that M_l is simply connected. Hence M_l is conformally equivalent either to $\mathbb{C}$ or to the unit disc Δ. We show that M_l cannot be conformally equivalent to $\mathbb{C}$. Indeed, by assumption we know the holomorphic maps $G^{(l)} : \mathbb{C} \to \mathbb{P}^n(\mathbb{C})$ omit more than $2n$ hyperplanes in general position in $\mathbb{P}^n(\mathbb{C})$, and by Corollary 4.3.5, $G^{(l)}$ must be constant. Thus $K_l \equiv 0$, which contradicts the fact that $K_l(p_l) = -\frac{1}{4}$. Therefore each M_l is conformally equivalent to the unit disc Δ. Without loss of generality, we assume that $M_l = \Delta$ and p_l is the origin, where p_l is given in (5.18).

By the assumption that the holomorphic maps $G^{(l)} : \Delta \to \mathbb{P}^n(\mathbb{C})$ omit more than $\frac{n+1}{2}(mn+2) \geq 2n$ Theorem 4.4.8 implies that $\{G^{(l)}\}$ is normal, that is, there exists a subsequence of holomorphic maps $\{G^{(l_i)}\}$ of $\{G^{(l)}\}$, still denoted by $\{G^{(l)}\}$, converging to a holomorphic map g uniformly on every compact subset of the unit disc Δ. Our goal is to derive a contradiction by proving that g is a nonconstant from one argument and g is constant from another argument. This will finish the proof.

We first argue that g is nonconstant. Assume that g is a constant. Then g maps the unit disc Δ onto a single point Q. Take a hyperplane H not containing the point Q, and let U_1 and U_2 be disjoint neighborhoods of H and Q, respectively. Choose a constant C as in Theorem 5.3.1 such that

$$|K(p)|^{\frac{1}{2}} d(p) \le C$$

for any map $G : M \to \mathbb{P}^n(\mathbb{C})$ that omits the neighborhood U of H. Note that the constant C is independent of M and G. Choose $r < 1$ such that the hyperbolic distance R from $z = 0$ to $|z| = r$ satisfies

$$R > 2C. \tag{5.20}$$

Since $G^{(l)}$ converges uniformly on $\Delta_r := \{z : |z| \le r\}$ to g, the image of Δ_r under $G^{(l)}$ lies in the neighborhood U_2 of Q for sufficiently large l. It follows that the image of Δ_r under $G^{(l)}$ omits the neighborhood U_1 of H for large l. Hence

$$|K_l(0)|^{\frac{1}{2}} d_l(r) \le C,$$

where $d_l(r)$ is the geodesic distance from the origin 0 to the boundary of Δ_r. Using the condition that $K_l(0) = -\frac{1}{4}$, for l large enough, we get

$$d_l(r) \le 2C. \tag{5.21}$$

Now we use Lemma 2.3.11 to derive a lower bound for $d_l(r)$. The surface M_l is a geodesic disk of radius $R_l(< +\infty)$. If we reparameterize it by letting $w = r_l z$, where the subset $\{w : |w| < r_l\}$ has a hyperbolic radius R_l, then the circle $|z| = r$ corresponds to $|w| = r_l r$. From (5.19) we know that $R_l \to \infty$ and thus $r_l \to 1$ as $l \to \infty$. Hence the hyperbolic radius of $|w| = r_l r$ tends to the hyperbolic radius of $|w| = r$, which is $R > 2C$ by the choice of R in (5.20). On the other hand, from Lemma 2.3.11 and $-1 \le K_l \le 0$ on M_l we get that the distance with the metric from the origin to any points on the circle $|z| = r$, or, equivalently, $|w| = r_l r$, is not less than the hyperbolic distance from the origin to any points on $|w| = r_l r$. Therefore $d_l(r) > 2C$, which contradicts (5.21). Hence we proved that the map g is not constant.

Finally, we show that g is a constant map. From (5.18), by Proposition 5.3.5 and the subsequent remark there is a subsequence of $\{g_j^{(l)}\}$, say itself, that converges to h_j uniformly on every compact subset of the unit disc Δ for each $0 \le j \le n$. Furthermore, $h_0, \ldots, h_n$ have no common zeros. So we get a holomorphic map $[h_0 : \cdots : h_n] : M \to \mathbb{P}^n(\mathbb{C})$. Obviously, $g = [h_0 : \cdots : h_n]$. We state that g is constant. Indeed, by Lemma 5.3.4 and (5.19) the conformal metric $ds^2 := \sum_{j=0}^n |h_j|^{2m} |dz|^2$ is complete on the unit disc Δ. Also, since $G^{(l)}$ omits the hyperplanes $\{H_j\}_{j=1}^q$ in $\mathbb{P}^n(\mathbb{C})$ located in general position with $q > \frac{n+1}{2}(mn+2)$, by Theorem 2.1.8 (Hurwitz's theorem), either the image g omits these hyperplanes $\{H_j\}_{j=1}^q$, or $g(\Delta) \subset \bigcap_{j=1}^t H_j$ for some $1 \le t \le q$. If g omits all these hyperplanes

$\{H_j\}_{j=1}^q$, then Corollary 5.2.2 implies that g must be constant. In the case where $g(\Delta) \subset \bigcap_{j=1}^t H_j = \mathbb{P}(V)$, where V is a subspace of $\mathbb{C}^{n+1}$ of dimension $n+1-t$, $g : \Delta \to \mathbb{P}(V)$ omits the hyperplanes $\{H_i \cap (\bigcap_{j=1}^t H_j)\}_{i=t+1}^q$ in $\mathbb{P}(V)$. Obviously, $H_{t+1} \cap (\bigcap_{j=1}^t H_j), \dots, H_q \cap (\bigcap_{j=1}^t H_j)$ are in general position. Since $q-t > \frac{n+1}{2}(mn+2)-t \geq \frac{n+1-t}{2}(m(n-t)+2)$, from Corollary 5.2.2 it again follows that g is a constant. This completes the proof of the theorem. □

5.4 Gauss maps of minimal surfaces in $\mathbb{R}^m$ with finite total curvature

Let M be a minimal surface immersed in $\mathbb{R}^m$, and let $\mathbf{x} : M \to \mathbb{R}^m$ be an immersion. Let $(U, \phi(u, v))$ be an isothermal local coordinate and denote $\mathbf{x}(u, v) = (\mathbf{x} \circ \phi^{-1})(u, v)$. By letting $z = u + iv$, M is a Riemann surface. The induced metric on M is given by $ds^2 = 2\|\phi\|^2|dz|^2$, where $\phi = (\phi_1, \dots, \phi_m) = \frac{\partial \mathbf{x}}{\partial z}$, and $\|\phi\|^2 = |\phi_1|^2 + \cdots + |\phi_m|^2$. The generalized Gauss map is $G = [\phi_1 : \cdots : \phi_m]$. We denote $\psi := \|\phi\|^2 \frac{\sqrt{-1}}{2\pi} dz \wedge d\bar{z}$. Note that the Gauss curvature is $K = -\frac{\mathrm{Ric}(\psi)}{\psi}$ by formula (2.8), so the (normalized) total curvature of M is

$$C(M) := \frac{1}{2\pi} \int_M K dA = -\int_M dd^c \log \|\phi\|^2 = -\int_M G^* \omega_{FS} \quad (\geq -\infty). \tag{5.22}$$

Throughout this section, we assume that $C(M)$ is finite, that is, $C(M) > -\infty$.

Theorem 5.4.1. *Let $\mathbf{x} : M \to \mathbb{R}^m$ be a complete regular minimal surface. Then the following conditions are equivalent: (a) $\mathbf{x} : M \to \mathbb{R}^m$ has finite total curvature $C(M)$; (b) There is a compact Riemann surface $\overline{M}$ such that the surface M canonically considered as a Riemann surface is biholomorphic to $\overline{M}$ with finitely many points $p_1, \dots, p_k$ excluded and each $\omega_i := \phi_i dz$, $1 \leq i \leq m$, can be extended to $\overline{M}$ as a meromorphic form; (c) The surface M is biholomorphic to $\overline{M}$ with finitely many points $p_1, \dots, p_k$ excluded, and the Gauss map $G : M \to \mathbb{P}^{m-1}(\mathbb{C})$ can be extended holomorphically to a holomorphic map of $\overline{M}$ into $\mathbb{P}^{m-1}(\mathbb{C})$.*

To prove the theorem, we need the following:

Lemma 5.4.2. *Let h be a nowhere zero holomorphic function on $D^*(a) := \{z \mid 0 < |z| < a\}$ $(0 < a < 1)$ and consider the metric $ds^2 = |h|^2|dz|^2$. If every piecewise smooth continuous curve in $D^*(a)$ converging to the origin has infinite length, then h has at worst a pole at the origin.*

Proof. Set $u := \log|h|$. By assumption, u is harmonic on $D^*(a)$. For $0 < r < a$, set

$$\beta \equiv \beta_1 + i\beta_2 := \frac{1}{2\pi i} \int_{|z|=r} \frac{\partial u}{\partial z} \quad (\beta_1, \beta_2 \in \mathbb{R}),$$

which does not depend on each r because $\partial u/\partial z$ is holomorphic. Let

$$g(z) := 2\int_{z_0}^{z}\left(\frac{\partial u}{\partial z}(\zeta) - \frac{\beta}{\zeta}\right)d\zeta$$

with arbitrary fixed $z_0 \in D^*(a)$. Note that g is a well-defined holomorphic function on $D^*(a)$. Then

$$\tilde{u}(z) := \operatorname{Re} g(z) = u(z) - (\beta_1 \log|z| - \beta_2 \arg z) - u(z_0).$$

Since $\tilde{u}$ and u are both single valued, we have $\beta_2 = 0$, and so $\beta = \beta_1$ is real. Consider the Laurent expansion

$$g(z) = g_1(z) + g_2(z) = \sum_{n>0}\frac{a_{-n}}{z^n} + \sum_{\geq 0} a_n z^n.$$

Set

$$u_1(z) := \operatorname{Re} g_1(z), \quad u_2(z) := \operatorname{Re} g_2(z).$$

Then

$$u = u_1 + u_2 + \beta \log|z| + u(z_0)$$

on $D^*(a)$. The function u_1 is extended to a harmonic function on $\mathbb{C}^*$, and the limit of u_1 exists as $z \to \infty$. On the other hand, u_2 is extendable to a harmonic function on $\{z \mid |z| < a\}$, and so it is bounded on $\{z \mid |z| < a'\}$ with $0 < a' < a$. Hence there exists a constant $A > 0$ such that

$$ds^2 = |h|^2|dz|^2 \leq Ae^{2u_1(z)}\left(\frac{1}{|z|}\right)^{2(N+2)}|dz|^2 \tag{5.23}$$

on $\{z \mid 0 < |z| < a'\}$, where N is a positive integer with $N+2 > -\beta$. Define the new metric

$$d\sigma^2 := Ae^{2u_1(z)}\left(\frac{1}{|z|}\right)^{2(N+2)}|dz|^2$$

on $\mathbb{C}^*$. By the change of variable $\zeta = 1/z$ we can rewrite

$$d\sigma^2 = Ae^{2u_1(1/\zeta)}|\zeta|^{2N}|d\zeta|^2$$

around $z = \infty$. Consider the holomorphic function

$$w(\zeta) := \int_0^{\zeta} \zeta^N e^{g_1(\zeta)} d\zeta$$

on $\mathbb{C}$. Let R_0 be the least upper bound of all R satisfying the condition that there exists a domain D such that $w(\zeta)$ maps D onto $D(R_0) := \{w \mid |w| < R_0\}$ and the map w is proper, that is, $w^{-1}(K) \cap D$ is compact for any compact set K in $D(R_0)$. The restriction of the map w to D gives a covering surface over $D(R_0)$, which is ramified with index N at the origin and unramified elsewhere. By the choice of R_0 we can find a smooth curve γ in D converging to the origin in the ζ-plane such that the image $w \circ \gamma$ is a segment from the origin to the boundary of $D(R_0)$. By the assumption,

$$\infty = \int_\gamma ds \le \int_\gamma d\sigma = \int_{w\circ\gamma} |dw| = R_0.$$

Hence we conclude that there is a covering isomorphism between the covering surface $w : D \to \mathbb{C}$ and $w^* : \mathbb{C} \to \mathbb{C}$ defined by $w^*(\zeta) = \zeta^{N+1}$. Since there is no proper domain of $\mathbb{C}$ that is biholomorphic with $\mathbb{C}$ itself, D coincides with the total complex plane $\mathbb{C}$. Moreover, since every biholomorphic automorphism of $\mathbb{C}$ that leaves the origin fixed is $\zeta \mapsto C\zeta$ for some constant C, it follows that g_1 is constant. By (5.23) we see that Lemma 5.4.2 holds. □

Proof of Theorem 5.4.1. Obviously, (b) implies (c) since $G = [\omega_1 : \cdots : \omega_m]$. Assume that M is biholomorphic to $\overline{M}$ with finitely many points $p_1, \ldots, p_k$ excluded and that the Gauss map $G : M \to \mathbb{P}^{m-1}(\mathbb{C})$ can be extended holomorphically to a holomorphic map $\tilde{G} : \overline{M} \to \mathbb{P}^{m-1}(\mathbb{C})$. Then

$$0 \ge C(M) = -\int_M \operatorname{Ric}(ds^2) \ge -\int_{\overline{M}} \operatorname{Ric}(ds^2) > -\infty,$$

which shows that M is of finite total curvature.

Now assume that M is of finite curvature. Then by Theorem 3.4.1 there exists a compact Riemann surface $\overline{M}$ such that M is biholomorphically equivalent to $\overline{M} - \{p_1, \ldots, p_k\}$. For each p_j, take a local holomorphic coordinate on a neighborhood U_j of p_j with $z(p_j) = \infty$ and $U_j := \triangle_{s,\infty} = \{z \mid s < |z| \le \infty\}$. Consider the characteristic function

$$T_G(r) = \int_s^r \frac{dt}{t} \int_{\triangle_{s,t}} G^*\omega_{FS}.$$

As we have noticed earlier,

$$C(M) = -\int_{\overline{M}} G^*\omega_{FS},$$

so the condition that $C(M)$ is finite implies that

$$\lim_{r\to\infty} \frac{T_G(r)}{\log r} < +\infty.$$

Then, by the big Picard theorem, G has a removable singularity at ∞. Take a new holomorphic coordinate $\zeta := 1/z$ on U_j and choose a new reduced representation $G = [g_1 : \cdots : g_m]$ around the origin. By shrinking U_j and changing the index $i = 1, \ldots, m$ if necessary we may assume that g_1 has no zero and $\nu_{g_1}(0) = \min_{1\le i\le m} \nu_{g_i}(0)$. Then each g_i/g_1 is holomorphic at $\zeta = 0$ and thus bounded around the origin. We can find a positive constant C such that

$$ds^2 = 2|\phi_1|^2\left(\sum_{i=1}^{m}\left|\frac{\phi_i}{\phi_1}\right|^2\right)|d\zeta| = 2|\phi_1|^2\left(\sum_{i=1}^{m}\left|\frac{g_i}{g_1}\right|^2\right)|d\zeta| \le C|\phi_1|^2|d\zeta|^2.$$

In this situation, we can apply Lemma 5.4.2 to the function $h := \phi_1$ to show that ϕ_1 has at worst a pole at the origin. This implies that each ω_i is extendable to $\overline{M}$ as a meromorphic form. Thus Theorem 5.4.1 is proved. □

Assume $C(M)$ is finite. Then by Theorem 5.4.1, $G : M \to \mathbb{P}^{m-1}(\mathbb{C})$ extends holomorphically to $\tilde{G} : \overline{M} \to \mathbb{P}^{m-1}(\mathbb{C})$. By the definition of $C(M)$ and by (2.17) we have

$$C_1(M) = -\deg(\tilde{G}). \tag{5.24}$$

The induced metric is $ds^2 = 2\sum_{k=1}^{n} |\phi_k(z)|^2|dz|^2 = 2\|\phi\|^2|dz|^2$ also extends to $\overline{M}$. However, it becomes a pseudo-metric on $\overline{M}$, which may have poles at $p_1, \ldots, p_k$. Write $ds^2 = |z|^{2\nu_j}h^2|dz|^2$ locally near p_j with $h(0) \ne 0$ and a local coordinate z such that $z(p_j) = 0$. Then, by exactly the same proof of (3.23), we have $\nu_j \le -2$ for $1 \le j \le k$. So by Theorem 2.4.1 (Gauss–Bonnet for pseudo-metric) and (5.24)

$$-\deg(G) = -\int_{\overline{M}} \mathrm{Ric}(\psi) = \chi(\overline{M}) + \sum_{j=1}^{k} \nu_j \le \chi(\overline{M}) - 2k, \tag{5.25}$$

where $\psi := \|\phi\|^2 \frac{\sqrt{-1}}{2\pi} dz \wedge d\bar{z}$ is the metric form associated with ds^2.

Theorem 5.4.3 ([68]). *Let M be a nonflat complete regular minimal surface in $\mathbb{R}^m$. Suppose that M is of finite total curvature. Then its generalized Gauss map $G : M \to \mathbb{P}^{m-1}(\mathbb{C})$ can fail to intersect at most $(m-1)(m+2)/2$ hyperplanes in general position in $\mathbb{P}^{m-1}(\mathbb{C})$.*

Proof. By Theorem 5.4.1, M is biholomorphic to $\overline{M} - \{p_1, \ldots, p_k\}$, and $G : M \to \mathbb{P}^{m-1}(\mathbb{C})$ extends to a holomorphic map from $\overline{M}$ into $\mathbb{P}^{m-1}(\mathbb{C})$. We still denote it by G. Since M is nonflat, G is not constant. So we assume that $G(\overline{M})$ is contained in $\mathbb{P}^l(\mathbb{C})$ but none of lower dimension with $l \ge 1$. Let $H_j, 1 \le j \le q$, be hyperplanes in general position in $\mathbb{P}^{m-1}(\mathbb{C})$ that do not intersect $G(M)$. By applying Theorem 2.5.6 to the map $G : \overline{M} \to \mathbb{P}^{m-1}(\mathbb{C})$ we have

$$(q - 2(m-1) + l - 1)d \le \frac{1}{2}l(2(m-1) - l + 1)\{2(g-1) + k\},$$

where $d = \deg(G)$, and g is the genus of $\overline{M}$. On the other hand, by (5.25)

$$d \geq 2(k + g - 1).$$

Hence we get

$$(q - (2m - l - 1))d \leq \frac{1}{2}l(2m - l - 1)(d - k),$$

or

$$\left[(q - (2m - l - 1)) - \frac{1}{2}l(2m - l - 1)\right]d < 0.$$

Hence

$$q < 2m - l - 1 + \frac{1}{2}l(2m - l - 1) = \frac{1}{2}(2m - l - 1)(2 + l) \leq m(m + 2)/2$$

since $1 \leq l \leq m - 1$. This proves our theorem. □

6 Nevanlinna theory on the open ball $B(R_0) \subset \mathbb{C}^n$ and the nonintegrated defect relation

In this chapter, we introduce the Nevanlinna theory on the open ball $B(R_0) \subset \mathbb{C}^n$. Following Xaiver's method [88], we also establish a nonintegrated defect relation for meromorphic maps of the complete Kähler manifold whose universal cover is the ball $B(R_0)$ with $0 < R_0 \leq \infty$ into $\mathbb{P}^N(\mathbb{C})$. See [26, 77, 84].

6.1 Meromorphic functions on $B(R_0) \subset \mathbb{C}^n$

First, we introduce some notations. We define $|z| = (|z_1|^2 + \cdots + |z_n|^2)^{1/2}$ for $z = (z_1, \ldots, z_n) \in \mathbb{C}^n$, $B(r) = \{z \in \mathbb{C}^n \mid |z| < r\}$ for $r > 0$ (sometimes, we denote it by $B_n(r)$ indicating its dependence on n), $B[r] = \{z \in \mathbb{C}^n \mid |z| \leq r\}$ (or $B_n[r]$), and $S(r) = \{z \in \mathbb{C}^n \mid |z| = r\}$ (or $S_n(r)$). The sphere $S(r)$ is considered as an analytic manifold oriented to the exterior of $B(r)$. Define $d^c = \frac{\sqrt{-1}}{4\pi}(\bar{\partial} - \partial)$. The pullback of the form

$$\sigma_n = d^c \log |z|^2 \wedge (dd^c \log |z|^2)^{n-1} \quad \text{on } \mathbb{C}^n - \{0\}$$

defines a positive measure on $S(r)$ with total measure 1. For $z \in \mathbb{C}^n - \{0\}$, let $\omega_n(z) = dd^c \log |z|^2$. For $z \in \mathbb{C}^n$, let $v_n(z) = dd^c|z|^2$ and $\rho_n(z) = (dd^c|z|^2)^n$. Then ρ_n is a Lebesgue measure on $\mathbb{C}^n$ such that $B(r)$ has measure r^{2n}.

A function h on an open domain $G \subset \mathbb{C}^n$ is said to be *holomorphic* if h is holomorphic in each variable. Let h be a nonconstant holomorphic function on an open domain $G \subset \mathbb{C}^n$. For a set $\alpha = (\alpha_1, \ldots, \alpha_n)$ of integers $\alpha_i \geq 0$, we set $|\alpha| = \alpha_1 + \cdots + \alpha_n$ and $D^\alpha h = D_1^{\alpha_1} \cdots D_n^{\alpha_n} h$, where $D_i h = (\partial/\partial z_i)h$, $i = 1, \ldots, n$. We define the divisor $\nu_h^0 : G \to \mathbb{Z}$ by

$$\nu_h^0(z) := \max\{k : D^\alpha h(z) = 0 \text{ for all } \alpha \text{ with } |\alpha| < k\} \quad (z \in G).$$

Alternatively, for each $a \in G$, expanding h as

$$h = \sum_m P_m(z - a)$$

with homogeneous polynomials P_m of degree m around a, we define

$$\nu_h^0(a) := \min\{m \mid P_m \not\equiv 0\}.$$

By a divisor on a domain G in $\mathbb{C}^n$ we mean a map ν of G into $\mathbb{Z}$ such that for each $z_0 \in G$, there are nonzero holomorphic functions h and g on a connected neighborhood $U(\subset G)$ of z_0 such that $\nu(z) = \nu_h^0(z) - \nu_g^0(z)$ for each $z \in U$ outside an analytic set of dimension $\leq n - 2$. We set

https://doi.org/10.1515/9783110989557-006

$$|\nu| = \overline{\{z \in G \mid \nu(x) \neq 0\}} \cap G,$$

which is a purely $(n-1)$-dimensional analytic set if $|\nu| \neq \emptyset$. Two divisors are regarded as the same if they are identical outside an analytic set of dimension $\leq n-2$. The *counting function* of a divisor ν on $B(R_0)$ $(0 < R_0 \leq +\infty)$ is defined as

$$n_\nu(r) = \frac{1}{r^{2n-2}} \int\limits_{|\nu| \cap B(r)} \nu \upsilon_n^{n-1}$$

if $n > 1$ and $n_\nu(r) = \sum_{z \in B(r)} \nu(z)$ if $n = 1$. The *valence function* of ν is defined as

$$N_\nu(r) = \int\limits_{r_0}^{r} \frac{n_\nu(t)}{t} dt$$

for $0 < r_0 < r < R_0$. It is independent of r_0 up to a bounded term.

Let φ be a nonconstant meromorphic function on $B(R_0) \subset \mathbb{C}^n$. For each $z_0 \in B(R_0)$, we choose nonzero holomorphic functions g and h on a neighborhood $U(\subset B(R_0))$ of z_0 such that $\varphi = \frac{g}{h}$ on U and $\dim(g^{-1}(0) \cup h^{-1}(0)) \leq n-2$. We define $\nu_\varphi^\infty := \nu_h$, $\nu_\varphi^a := \nu_{g-ah}$ for $a \in \mathbb{C}$, and $\nu_\varphi = \nu_\varphi^0 - \nu_\varphi^\infty$, which are independent of the choices of h and g and so are globally well-defined on $B(R_0)$. For brevity, set $N_\varphi(r, a) := N_{\nu_\varphi^a}(r)$ for $a \in \mathbb{C} \cup \{\infty\}$. We have the Jensen formula (see [83], p. 248): *For $0 < r_0 < r < R_0$, we have*

$$N_\varphi(r, 0) - N_\varphi(r, \infty) = \int\limits_{S(r)} \log|\varphi| \sigma_n + O(1).$$

The proximity function $m_\varphi(r, a)$ is defined by

$$m_\varphi(r, a) = \int\limits_{S(r)} \log^+ \frac{1}{|\varphi - a|} \sigma_n \quad \text{if } a \neq \infty \tag{6.1}$$

and

$$m_\varphi(r, \infty) = \int\limits_{S(r)} \log^+ |\varphi| \sigma_n. \tag{6.2}$$

The *characteristic function* of φ is defined as

$$T_\varphi(r) := m_\varphi(r, \infty) + N_\varphi(r, \infty).$$

Next, we extend the logarithmic derivative lemma in Chapter 4 to meromorphic functions on $\mathbb{C}^n$, following the methods of Biancofiore and Stoll [7] and Fujimoto [26].

Lemma 6.1.1. *Let $r > 0$, and let h be a function on $S_n(r)$ such that $h\sigma_n$ is integrable over $S_n(r)$. Let $p(w) = \sqrt{r^2 - |w|^2}$. Then*

$$\int_{S_n(r)} h\sigma_n = \frac{1}{r^{2n-2}} \int_{B_{n-1}[r]} \Big(\int_{S_1(p(w))} h(w,\zeta)\sigma_1(\zeta) \Big) \rho_{n-1}(w).$$

Proof. Define $E = \{(z_1, \ldots, z_n) \in S_n(r) \mid 0 \le z_n \in \mathbb{R}\}$ and $F = S_n(r) - E$. Then E is a closed subset of $S_n(r)$ of zero $(2n-1)$-dimensional Hausdorff measure. A bijective map

$$g : \mathbb{R}(0, 2\pi) \times B_{n-1}(r) \longrightarrow F$$

of class C^∞ is defined by $g(\phi, w) = (w, p(w)e^{i\phi})$ for all $\phi \in \mathbb{R}(0, 2\pi)$ and $w \in B_{n-1}(r)$. Let ϕ be the variable in $\mathbb{R}(0, 2\pi)$, and let $w_1, \ldots, w_{n-1}$ be the complex variables on $\mathbb{C}^{n-1}$ with $w_j = x_j + y_j\sqrt{-1}$. The partial derivatives $g_\phi, g_{x_1}, g_{y_1}, \ldots, g_{x_{n-1}}, g_{y_{n-1}}$ are pointwise linearly independent over $\mathbb{R}$, and g is perpendicular to these derivatives. Here $g(\phi, w)$ points in the direction of $S_{n-1}(r)$ at $g(\phi, w)$. Since

$$\det(g, g_\phi, g_{x_1}, g_{y_1}, \ldots, g_{x_{n-1}}, g_{y_{n-1}}) = r^2 > 0,$$

the map g is an orientation-preserving diffeomorphism. We will compute $g^*(\sigma_n)$. For $z = (z_1, \ldots, z_n) \in \mathbb{C}^n$ and $w = (w_1, \ldots, w_{n-1}) \in \mathbb{C}^{n-1}$, we have

$$\begin{aligned} g^*(d^c|z|^2) &= d^c|w|^2 + (1/2\pi)(r^2 - |w|^2)d\phi, \\ g^*(dd^c|z|^2) &= dd^c|w|^2 - (1/2\pi)d|w|^2 \wedge d\phi, \\ g^*(dd^c|z|^2)^{n-1} &= (dd^c|w|^2)^{n-1} - \frac{n-1}{2\pi} d|w|^2 \wedge d\phi \wedge (dd^c|w|^2)^{n-2}, \\ g^*(d^c|z|^2 \wedge (dd^c|z|^2)^{n-1}) &= (1/2\pi)(r^2 - |w|^2)d\phi \wedge (dd^c|w|^2)^{n-1} \\ &\quad + \frac{n-1}{2\pi} d\phi \wedge d|w|^2 \wedge d^c|w|^2 \wedge (dd^c|w|^2)^{n-2}. \end{aligned}$$

It is easy to check that $|w|^2(dd^c|w|^2)^{n-1} = (n-1)d|w|^2 \wedge d^c|w|^2 \wedge (dd^c|w|^2)^{n-2}$, and hence

$$g^*(d^c|z| \wedge (dd^c|z|^2)^{n-1}) = (1/2\pi)r^2 d\phi \wedge (dd^c|w|^2)^{n-1}.$$

Now $|g|^2 = r^2$ and $(dd^c|w|^2)^{n-1} = \rho_{n-1}$, and we obtain

$$\begin{aligned} g^*(\sigma_n) &= g^*(d^c \log|z|^2 \wedge (dd^c \log|z|^2)^{n-1}) = g^*(r^{-2n} d^c|z|^2 \wedge (dd^c|z|^2)^{n-1}) \\ &= (1/2\pi)r^{2-2n} d\phi \wedge \rho_{n-1}. \end{aligned}$$

Fubini's theorem implies that

$$\int_{S_n(r)} h\sigma_n = r^{2-2n} \int_{B_{n-1}[r]} \frac{1}{2\pi} \int_0^{2\pi} h(w, p(w)e^{i\phi})d\phi \wedge \rho_{n-1}(w). \qquad \square$$

Lemma 6.1.2. *Let ϕ be a nonconstant meromorphic function on $B(R_0)$ with $0 < R_0 \leq \infty$, and let $a \in \mathbb{C} \cup \{\infty\}$. Then, for $0 < r < R_0$,*

$$\frac{1}{r^{2n-2}} \int_{B_{n-1}[r]} n_{\phi_{[w]}}(\sqrt{r^2 - |w|^2}, a)\rho_{n-1}(w) \leq n_\phi(r, a),$$

where $\phi_{[w]} = \phi(w, z)$ for $w \in \mathbb{C}^{n-1}$ and $z \in \mathbb{C}$.

Proof. For each $j \in \mathbb{N}[1, n]$, define $\pi_j : \mathbb{C}^n \to \mathbb{C}^{n-1}$ by

$$\pi_j(z_1, \dots, z_n) = (z_1, \dots, z_{j-1}, z_{j+1}, \dots, z_n).$$

Then

$$v_n^{n-1} = \sum_{j=1}^{n} \pi_j^*(v_{n-1}^{n-1}) \geq \pi_n^*(\rho_{n-1}).$$

Define $A = B_n(r) \cap \operatorname{supp} \nu_\phi^a$. If A is empty, then the lemma is trivial. Assume that A is not empty. Then A is a pure $(n-1)$-dimensional subset of $B_n(r)$. Denote $\pi = \pi_n|_A$. Then $\pi(A) \subset B_{n-1}(r)$. The set $E = \{z \in A \mid \operatorname{rank}_z \pi < n-1\}$ is analytic in $B_n(r)$. Let E_0 be the union of all $(n-1)$-dimensional branches of E, and let E_1 be the union of all other branches of E. Then E_0 and E_1 are analytic in $B_n(r)$ with $\dim E_1 \leq n-2$ and $\dim E_0 = n-1$ if $E_0 \neq \emptyset$. Also, $E_0' = \pi(E_0)$ is analytic in $B_{n-1}(r)$ with $E_0 = \pi_n^{-1}(E_0') \cap B_n(r)$. Therefore $\pi^*(\rho_{n-1}) = 0$ on E_0. The complement $A_0 = A - E_0$ is open in A, and $A_0 \cap E_1$ is thin analytic in A_0 if $A_0 \neq \emptyset$. Then $A_1 = A_0 - E_1 = A - E$ is open in A. There exists a thin analytic subset E_2 of A_1 such that π is locally biholomorphic on $A_2 = A_1 - E_2$. Then $F = \pi(E_0 \cup E_1 \cup E_2)$ has a zero measure in $\mathbb{C}^{n-1}$. Also, $B_2 = \pi(A_2)$ is open in $\mathbb{C}^{n-1}$. Here $B = \pi(A) = B_2 \cup F$ and $B_3 = B_2 - F$ differ from B_2 by sets of zero measure.

The intersection $S_n(r) \cap \operatorname{supp} \nu_\phi^a$ has a $(2n-2)$-dimensional Hasudorff measure of zero. If $w \in B_{n-1}(r)$, then

$$\pi^{-1}(w) = (\{w\} \times B_1(\sqrt{r^2 - |w|^2})) \cap A,$$

and for almost all $w \in B_{n-1}(r)$, we have

$$\pi^{-1}(w) = (\{w\} \times B_1[\sqrt{r^2 - |w|^2}]) \cap \operatorname{supp} \nu_\phi^a.$$

For all $w \in B_{n-1}(r) - B$, we have $\pi^{-1}(w) = \emptyset$, and hence $n_{f_{[w]}}(\sqrt{r^2 - |w|^2}, a) = 0$ for almost all $w \in B_{n-1}(r) - B$. If $w \in B_3$, then

$$\pi^{-1}(w) = (\{w\} \times B_1(\sqrt{r^2 - |w|^2})) \cap A.$$

Take $P = (b, c) \in A_2$ with $b = \pi(P) \in B_{n-1}(r)$ and $c \in \mathbb{C}$. There exist open connected neighborhoods V of b in $B_{n-1}(r)$, W of c, and $U = V \times W$ of P in $B_n(r)$ with $U \cap A_2 = U \cap A$

such that $\pi : U \cap A \to V$ is biholomorphic. Let λ be the inverse map. Then there exists a holomorphic function $h : V \to W$ such that $\lambda(w) = (w, h(w))$ for all $w \in V$. The multiplicity $\nu_\phi^a(w,z) = q$ is constant for all $(w,z) \in U \cap A$. There exists a holomorphic function $H : U \to \mathbb{C} - \{0\}$ such that

$$\phi(w,z) = a + (z - h(w))^q H(w,z)$$

for all $(w,z) \in U$. Therefore $\nu_\phi^a(w,z) = q = \nu_{\phi_{[w]}}^a(z)$ for all $(w,z) \in U \cap A$. All together, we obtain

$$\begin{aligned} r^{2n-2} n_\phi(r,a) &= \int_A \nu_\phi^a \upsilon_n^{n-1} \geq \int_A \nu_\phi^a \pi^*(\rho_{n-1}) = \int_{A_0} \nu_\phi^a \pi^*(\rho_{n-1}) \\ &= \int_{A_2} \nu \pi^*(\rho_{n-1}) = \int_{B_2} \sum_{(w,z)\in A_2} \nu_\phi^a(w,z) \rho_{n-1}(w) \\ &= \int_{B_3} \Big(\sum_{z \in B_1(\sqrt{r^2 - |w|^2})} \nu_{\phi_{[w]}}^a(z) \Big) \rho_{n-1}(w) \\ &= \int_{B_3} n_{\phi_{[w]}}(\sqrt{r^2 - |w|^2}, a) \rho_{n-1}(w) \\ &= \int_{B} n_{\phi_{[w]}}(\sqrt{r^2 - |w|^2}, a) \rho_{n-1}(w) \\ &= \int_{B_{n-1}(r)} n_{\phi_{[w]}}(\sqrt{r^2 - |w|^2}, a) \rho_{n-1}(w). \end{aligned}$$

□

Theorem 6.1.3. *Let ϕ be nonzero a meromorphic function on $B(R_0) \subset \mathbb{C}^n$, $0 < R_0 \leq \infty$, let $\alpha = (\alpha_1, \dots, \alpha_n) \neq (0, \dots, 0)$, $0 < r_0 < R_0$, and let p, p' be positive numbers such that $0 < p|\alpha| < p' < 1$. Then, for $r_0 < r < R < R_0$,*

$$\int_{S(r)} \left| z^\alpha \left(\frac{D^\alpha \phi}{\phi} \right)(z) \right|^p \sigma_n(z) \leq K \left(\frac{R^{2n-1}}{R - r} T_\phi(R) \right)^{p'},$$

where K is a constant not depending on r and R, and $z^\alpha := z_1^{\alpha_1} \cdots z_n^{\alpha_n}$ for $z = (z_1, \dots, z_n)$ and $\alpha = (\alpha_1, \dots, \alpha_n)$.

Before proving this, we give the following corollary to Theorem 6.1.3, which is essentially the same as the lemma of the logarithmic derivative in several complex variables given by Vitter [86].

Corollary 6.1.4. *Let $\alpha = (\alpha_1, \dots, \alpha_n) \neq (0, \dots, 0), 0 < r_0 < R_0$. Then, for $r_0 < r < R < R_0$,*

$$\int_{S(r)} \log^+ |(D^\alpha \phi/\phi)(z)| \sigma_n(z) \leq K \log^+\left(\frac{R^{2n-1}}{R-r} T_\phi(R)\right).$$

Proof of Theorem 6.1.3. First, write $\phi = \phi(\eta, \zeta)$ with $\zeta \in \mathbb{C}$. From (4.12) we have, with $\tilde{p} = p/p'$,

$$\begin{aligned}
&\int_{|\zeta| = \sqrt{r^2 - |\eta|^2}} \left|\zeta\left(\frac{\partial \phi}{\partial \zeta}/\phi\right)(\eta, \zeta)\right|^{\tilde{p}} \sigma_1(\zeta) \\
&\leq \left(\frac{\rho}{\rho - r} \int_{|\zeta| = \sqrt{r^2 - \eta^2}} |\log|\phi(\eta, \zeta)|| \sigma_1(\zeta)\right)^{\tilde{p}} \\
&\quad + K(n_\phi(\sqrt{\rho^2 - |\eta|^2}, 0) + n_\phi(\sqrt{\rho^2 - |\eta|^2}, \infty)).
\end{aligned} \tag{6.3}$$

We prove Theorem 6.1.3 by induction on $|\alpha|$. We first consider the case where $|\alpha| = 1$. Without loss of generality, we assume that $D^\alpha = D_n$. Set $\rho = (r + R)/2$. Since each pole of $D_n\phi/\phi$ is of order ≤ 1, $|z_n(D_n\phi/\phi)(z)|^{\tilde{p}}$ is integrable on $S(r)$. By Lemma 6.1.1, Lemma 6.1.2, (6.3), and the Hölder inequality we have

$$\begin{aligned}
&\int_{S(r)} |z_n(D_n\phi/\phi)(z)|^{\tilde{p}} \sigma_n(z) \\
&= \frac{1}{r^{2n-2}} \int_{|\eta| \leq r} v_{n-1}^{n-1}(\eta) \int_{|\zeta| = \sqrt{r^2 - \eta^2}} |\zeta(D_n\phi/\phi)(\eta, \zeta)|^{\tilde{p}} \sigma_1(\zeta) \\
&\leq \frac{1}{r^{2n-2}} \left(\frac{\rho}{\rho - r}\right)^{\tilde{p}} \left(\int_{|\eta| \leq r} v_{n-1}^{n-1}(\eta)\right)^{1-\tilde{p}} \\
&\quad \times \left(\int_{|\eta| \leq r} v_{n-1}^{n-1}(\eta) \int_{|\zeta| = \sqrt{\rho^2 - \eta^2}} |\log|\phi(\eta, \zeta)|| \sigma_1(\zeta)\right)^{\tilde{p}} \\
&\quad + \frac{K}{r^{2n-2}} \int_{|\eta| \leq r} \{n_{\phi|\eta}(\sqrt{\rho^2 - |\eta|^2}), 0) + n_{\phi|\eta}(\sqrt{\rho^2 - |\eta|^2}), \infty)\} v_{n-1}^{n-1}(\eta) \\
&\leq \left(\frac{\rho}{\rho - r} \int_{S(\rho)} |\log|\phi(\eta, \zeta)|| \sigma_n\right)^{\tilde{p}} + K\left(\frac{\rho}{r}\right)^{2n-2} \{n_{\phi|\eta}(\rho, 0) + n_{\phi|\eta}(\rho, \infty)\}.
\end{aligned}$$

Therefore

$$\begin{aligned}\int_{S(r)} |z_n(D_n\phi/\phi)(z)|^p \sigma_n(z) &\le \Big(\int_{S(r)} |z_n(D_n\phi/\phi)(z)|^{\tilde{p}} \sigma_n(z)\Big)^{p'} \\ &\le \Big(\frac{\rho}{\rho - r}\int_{S(\rho)} |\log|\phi||\sigma_n\Big)^{\tilde{p}p'} \\ &\quad + K\Big(\frac{\rho}{r}\Big)^{(2n-2)p'} (n_{\phi|\eta}(\rho,0)^{p'} + n_{\phi|\eta}(\rho,\infty)^{p'}).\end{aligned}$$

By the first main theorem

$$\int_{S(r)} |\log|\phi||\sigma_n \le 2T_\phi(r) + K,$$

and by (4.13) with $a = 0$ and ∞

$$n_\phi(\rho, a) \le \frac{2R}{R - r}(T_\phi(R) + K),$$

where K stands for a constant, which may be different each time. Thus

$$\begin{aligned}&\int_{S(r)} |z_n(D_n\phi/\phi)(z)|^p \sigma_n(z) \\ &\quad \le \Big(\frac{2R}{R-r}\int_{S(\rho)} |\log|\phi||\sigma_n\Big)^p + K\Big(\frac{4R^{2n-1}}{R-r}(T_\phi(R) + \tilde{K})\Big)^{p'} \\ &\quad \le K\Big(\frac{R^{2n-1}}{R-r}T_\phi(R)\Big)^{p'}. \end{aligned} \tag{6.4}$$

This proves the theorem for the case $D^\alpha = D_n$. We now consider the general case. Assume that the theorem holds for the case $|\alpha| \le \kappa$. Then $D^\alpha\phi/\phi = (D_i\phi/\phi)(D^{\alpha'}(D_i\phi)/D_i\phi)$, $z^\alpha = z_i z^{\alpha'}$, and $|\alpha|p = (|\alpha'| + 1)p < p' < 1$. Set $p_1 := 1/(|\alpha'| + 1)$ and $p_2 := |\alpha'|/(|\alpha'| + 1)$. By the Hölder inequality, the induction hypothesis, and (6.4) we have

$$\begin{aligned}&\int_{S(r)} |z^\alpha(D^\alpha\phi/\phi)(z)|^p \sigma_n(z) \\ &\quad \le \Big(\int_{S(r)} |z_i(D_i\phi/\phi)(z)|^{p/p_1}\sigma_n(z)\Big)^{p_1}\Big(\int_{S(r)} |z^{\alpha'}(D^{\alpha'}(D_i\phi)/D_i\phi)(z)|^{p/p_2}\sigma_n(z)\Big)^{p_2} \\ &\quad \le K\Big(\frac{R^{2n-1}}{R-r}T_\phi(R)\Big)^{p'p_1}\Big(\frac{R^{2n-1}}{R-r}T_{D_i\phi}(R)\Big)^{p'p_2}. \end{aligned} \tag{6.5}$$

Notice that

$$T_{D_i\phi}(r) \le T_{D_i\phi/\phi}(r) + T_\phi(r) + K$$
$$\le \int_{S(r)} \log^+ |D_i\phi/\phi|\sigma_n + N_{D_i\phi/\phi}(r,\infty) + T_\phi(r) + K.$$

Since

$$N_{D_i\phi/\phi}(r,\infty) \le N_\phi(r,\infty) + N_\phi(r,0) \le 2T_\phi(r) + K$$

and

$$\int_{S(r)} \log^+ |(D_i\phi/\phi)(z)|\sigma_n \le K \log^+\left(\frac{R^{2n-1}}{R-r} T_\phi(R)\right)$$

by (6.4), we have

$$T_{D_i\phi}(r) \le 3T_\phi(r) + K\log^+\left(\frac{R^{2n-1}}{R-r} T_\phi(R)\right) \quad \text{for } i = 1, 2, \ldots, n. \tag{6.6}$$

For any $\epsilon > 0$, there exists a positive constant K_ϵ such that

$$\log^+\left(\frac{R^{2n-1}}{R-r} T_\phi(R)\right) \le K_\epsilon\left(\frac{R^{2n-1}}{R-r} T_\phi(R)\right)^\epsilon. \tag{6.7}$$

Thus, for slightly smaller p' if necessary, by combining (6.5), (6.6), and (6.7) we can conclude

$$\int_{S(r)} \left|z^\alpha (D^\alpha\phi/\phi)(z)\right|^p \sigma_n(z) \le K\left(\frac{R^{2n-1}}{R-r} T_\phi(R)\right)^{p'}.$$

This completes the proof of the theorem. □

6.2 Nevanlinna theory on $B(R_0) \subset \mathbb{C}^n$

Let $f : B(R_0) \subset \mathbb{C}^n \to \mathbb{P}^N(\mathbb{C})$ be a meromorphic map with $0 < R_0 \le \infty$. We take holomorphic functions $f_0, f_1, \ldots, f_N$ such that $I_f := \{z \in B(R_0) \mid f_0(z) = \cdots = f_N(z) = 0\}$ is of dimension at most $n-2$ and $f(z) = [f_0(z) : \cdots : f_N(z)]$ on $B(R_0) - I_f$ in terms of homogeneous coordinates $[w_0 : \cdots : w_N]$ on $\mathbb{P}^N(\mathbb{C})$. We call such a representation $\mathbf{f} = (f_0, \ldots, f_N)$ a reduced representation of f. The pullback of the normalized Fubini–Study metric form Ω on $\mathbb{P}^N(\mathbb{C})$ by f is given by

$$\Omega_f = dd^c \log \|f\|^2,$$

where $\|f\| := \max_{0\le j\le N} |f_j|$. The *characteristic function* of f is defined as

$$T_f(r) = \int_{r_0}^{r} \frac{dt}{t^{2n-1}} \int_{B(t)} \Omega_f \wedge v_n^{n-1}$$

for $0 < r_0 < r < R_0$, which is independent of r_0 up to a bounded term. Let H be a hyperplane defined by $\mathbf{a}$, that is,

$$H = \{[w_0 : \cdots : w_N] \mid a_0 w_0 + \cdots + a_N w_N = 0\} \subset \mathbb{P}^N(\mathbb{C}),$$

where $\mathbf{a} = (a_0, \ldots, a_N)$. The proximity function $m_f(r, H)$ for $0 < r < R_0$ is defined by

$$m_f(r,H) = \int_{S(r)} \lambda_H(f)\sigma_n, \tag{6.8}$$

where the *Weil function* $\lambda_H(f)$ *of* f *with respect to the hyperplane* H is

$$\lambda_H(f)(z) = \log \frac{\|\mathbf{f}(z)\|\|\mathbf{a}\|}{|< \mathbf{f}(z), \mathbf{a} >|}.$$

The first main theorem states (see [83], pp. 251–255) that

$$T_f(r) = m_f(r,H) + N_f(r,H) + O(1)$$

for $0 < r < R_0$. Next, we prove the second main theorem. We say a map $f : B(R_0) \subset \mathbb{C}^n \to \mathbb{P}^N(\mathbb{C})$ is *linearly nondegenerate* if $f(B(R_0)) \not\subset H$ for every hyperplane H in $\mathbb{P}^N(\mathbb{C})$ or, equivalently, $f_0, \ldots, f_N$ are linearly independent, where $\mathbf{f} = (f_0, \ldots, f_N)$ is a reduced representation of f. Recall that for a holomorphic function h on an open domain $G \subset \mathbb{C}^n$ and a set $\alpha = (\alpha_1, \ldots, \alpha_n)$ of integers $\alpha_i \geq 0$, we denote $|\alpha| := \alpha_1 + \cdots + \alpha_n$ and $D^\alpha h := D_1^{\alpha_1} \cdots D_n^{\alpha_n} h$, where $D_i h = (\partial/\partial z_i)h$ for $i = 1, \ldots, n$. Take an arbitrary set $\alpha^i = (\alpha_{i1}, \ldots, \alpha_{in}), 1 \leq i \leq N+1$, of nonnegative integers. We define the generalized Wronskian of f as

$$W_{\alpha^1 \ldots \alpha^{N+1}}(f) = \det(D^{\alpha^i}\mathbf{f} : 1 \leq i \leq N+1),$$

where $\mathbf{f} = (f_0, \ldots, f_N)$ is a reduced representation of f, and $D^\alpha \mathbf{f} = (D^\alpha f_0, \ldots, D^\alpha f_N)$. We prove the following theorem.

Theorem 6.2.1. *Let $f : B(R_0) \to \mathbb{P}^N(\mathbb{C})$ be a linearly nondegenerate meromorphic map. Then there exist $\alpha^j = (\alpha_{j1}, \ldots, \alpha_{jn})$ with integers $\alpha_{ji} \geq 0$, $|\alpha^j| \leq N$ for $1 \leq j \leq N+1$, and $|\alpha^1| + \cdots + |\alpha^{N+1}| \leq N(N+1)/2$, such that $W_{\alpha^1 \ldots \alpha^{N+1}}(f) \not\equiv 0$.*

To prove Theorem 6.2.1, we denote by $\mathcal{M}_p$ the field of all germs of meromorphic functions at a point $p \in B(R_0)$ and by $\mathcal{F}^\kappa$ the $\mathcal{M}_p$-submodule of $\mathcal{M}_p^{N+1}$ generated by $\{D^\alpha \mathbf{f} : |\alpha| \leq \kappa\}$, where $D^\alpha \mathbf{f} = (D^\alpha f_0, \ldots, D^\alpha f_N) \in \mathcal{M}_p^{N+1}$. Set $\mathcal{F} = \bigcup_\kappa \mathcal{F}^\kappa$.

Proposition 6.2.2. *The set $\mathcal{F}^\kappa$ does not depend on the choice of holomorphic local coordinates $(z_1, \ldots, z_n)$ and reduced representation $\mathbf{f} = (f_0, \ldots, f_N)$ of f.*

Proof. We will show this by induction on κ. For the case $\kappa = 0$, we have nothing to prove. Assume that the statement holds for $|\alpha| \le \kappa$. We consider another system of holomorphic local coordinates $u = (u_1, \ldots, u_n)$. We denote

$$D_{i,u} = \frac{\partial}{\partial u_i}, \quad D^{\alpha,u} = \frac{\partial^{|\alpha|}}{\partial u_1^{\alpha_1} \cdots \partial u_n^{\alpha_n}}$$

for $\alpha = (\alpha_1, \ldots, \alpha_n)$. For an arbitrary α with $|\alpha| = \kappa + 1$, we write $D^\alpha = D_i D^{\alpha'}$, where $1 \le i \le n$ and $|\alpha'| = \kappa$. By the induction hypothesis we can write

$$D^{\alpha',u}\mathbf{f} = \sum_{|\beta| \le \kappa} h_\beta D^\beta \mathbf{f}$$

with some $h_\beta \in \mathcal{M}_p$. Then

$$D^{\alpha,u}\mathbf{f} = D_{i,u}(D^{\alpha',u}\mathbf{f}) = \sum_{|\beta| \le \kappa} D_{i,u} h_\beta D^\beta \mathbf{f} + \sum_{|\beta| \le \kappa} h_\beta \sum_{j=1}^{n} \frac{\partial z_j}{\partial u_i} D_j D^\beta \mathbf{f} \in \mathcal{F}^{\kappa+1}.$$

This proves that the set $\mathcal{F}^{\kappa+1}$ does not depend on a particular choice of a system of holomorphic local coordinates.

Next, we consider another reduced representation $\tilde{f} = (\tilde{f}_0, \ldots, \tilde{f}_N)$ of f. Then $h_i := \tilde{f}_i / f_i$, $i = 0, \ldots, N$, are nowhere zero holomorphic functions. Let $D^\alpha = D_i D^{\alpha'}$ for $|\alpha'| = \kappa$. By the induction hypothesis we can write

$$D^{\alpha'}\tilde{f} = \sum_{|\beta| \le \kappa} g_\beta D^\beta \mathbf{f}$$

with some $g_\beta \in \mathcal{M}_p$. Therefore

$$D^\alpha \tilde{f} = D_i(D^{\alpha'}\tilde{f}) = \sum_{|\beta| \le \kappa} D_i g_\beta D^\beta \mathbf{f} + \sum_{|\beta| \le \kappa} g_\beta D_i D^\beta \mathbf{f} \in \mathcal{F}^{\kappa+1}.$$

Hence the set $\mathcal{F}^{\kappa+1}$ does not depend on a particular choice of reduced representations.

The proposition is thus proved by induction. □

Proposition 6.2.3. *The map $f : B(R_0) \subset \mathbb{C}^n \to \mathbb{P}^N(\mathbb{C})$ is linearly nondegenerate if and only if $\mathcal{F} = \mathcal{M}_p^{N+1}$ for some point p or, equivalently, $l(\kappa_0) = N + 1$ for some κ_0, where $l(\kappa) = \dim_{\mathcal{M}_p} \mathcal{F}^\kappa$.*

Proof. Assume that f is degenerate. Then

$$a_0 f_0 + \cdots + a_N f_N \equiv 0$$

for some $(a_0, \dots, a_n) \neq (0, \dots, 0)$. Therefore

$$a_0 D^\alpha f_0 + \cdots + a_N D^\alpha f_N \equiv 0$$

for all α. Therefore $l(\kappa) = \operatorname{rank}_{\mathcal{M}_p}\{D^\alpha \mathbf{f} : |\alpha| \leq \kappa\} < N+1$ for every $\kappa = 1, 2, \dots$.

Conversely, assume that $\max_\kappa l(\kappa) < N+1$. Then there exists $(\phi_0, \dots, \phi_N) \neq (0, \dots, 0)$ in $\mathcal{M}_p^{N+1}$ such that

$$\phi_0 D^\alpha f_0 + \cdots + \phi_N D^\alpha f_N \equiv 0$$

for all α. Take a point $q \in M$ sufficiently near p such that $\phi_0, \dots, \phi_N$ are holomorphic in a neighborhood of q and $(\phi_0(q), \dots, \phi_N(q)) \neq (0, \dots, 0)$. Set

$$\Psi(z) := \phi_0(q) f_0(z) + \cdots + \phi_N(q) f_N(z)$$

on U. Then

$$(D^\alpha \Psi)(q) := \phi_0(q) D^\alpha f_0(q) + \cdots + \phi_N(q) D^\alpha f_N(q) = 0$$

for all α. Hence $\Psi \equiv 0$ is constant. This shows that f is degenerate. □

Lemma 6.2.4. *If $\mathcal{F}^\kappa = \mathcal{F}^{\kappa+1}$ for a positive integer κ, then $\mathcal{F} = \mathcal{F}^\kappa$.*

Proof. We show that $\mathcal{F}^{\kappa'} \subset \mathcal{F}^\kappa$ for every $\kappa' (> \kappa)$ by induction. There is nothing to prove in the case $\kappa' = \kappa + 1$. Assume that $\mathcal{F}^{\kappa'} \subset \mathcal{F}^\kappa$ for $\kappa' > \kappa$. Take α with $|\alpha| = \kappa' + 1$ and write $D^\alpha = D_i D^{\alpha'}$, where $1 \leq i \leq n$ and $|\alpha'| = \kappa$. By the induction hypothesis we can write

$$D^{\alpha'} \mathbf{f} = \sum_{|\beta| \leq \kappa} \phi_\beta D^\beta \mathbf{f}$$

with some $\phi_\beta \in \mathcal{M}_p$. Then

$$D^\alpha \mathbf{f} = D_i(D^{\alpha',u} \mathbf{f}) = \sum_{|\beta| \leq \kappa} D_i \phi_\beta D^\beta \mathbf{f} + \sum_{|\beta| \leq \kappa} \phi_\beta D_j D^\beta \mathbf{f} \in \mathcal{F}^{\kappa+1} = \mathcal{F}^\kappa.$$

Thus $\mathcal{F}^{\kappa'+1} \subset \mathcal{F}^\kappa$, and the lemma is proved by induction. □

Proof of Theorem 6.2.1. By Proposition 6.2.3 there is at least one κ with $l(\kappa) = N + 1$. Set

$$\kappa_0 = \min\{\kappa : l(\kappa) = N + 1\}.$$

We claim that $\kappa_0 \leq N$. Indeed, Lemma 6.2.4 implies that

$$1 = l(0) < l(1) < \cdots < l(\kappa_0) = N + 1.$$

Therefore

$$N + 1 = l(\kappa_0) = \sum_{\kappa=1}^{\kappa_0} (l(\kappa) - l(\kappa - 1)) + l(0) \geq \kappa_0 + 1.$$

This gives $\kappa_0 \leq N$. Thus our claim is proved.

Next, note that $W_{\alpha^1 \cdots \alpha^{N+1}}(f) \not\equiv 0$ if and only if $D^{\alpha_1}\mathbf{f}, \dots, D^{\alpha_{N+1}}\mathbf{f}$ is a basis of $\mathcal{F}$ as an $\mathcal{M}_p$-module. We choose a basis of $\mathcal{F}$ according to the following filtration by using Lemma 6.2.4 and the fact that $l(\kappa_0) = N + 1$:

$$\mathcal{F}^1 \subset \mathcal{F}^1 \subset \cdots \subset \mathcal{F}^{\kappa_0},$$

that is, $\{D^{\alpha_1}\mathbf{f}, \dots, D^{\alpha_{N+1}}\mathbf{f}\}$ is a basis of $\mathcal{F}$ such that $\{D^{\alpha_1}\mathbf{f}, \dots, D^{\alpha_{l(\kappa)}}\mathbf{f}\}$ is a basis of $\mathcal{F}^\kappa$ for $\kappa = 1, 2, \dots, \kappa_0$. Such a basis is called an *admissible* basis of $\mathcal{F}$. With this choice, we have

$$\begin{aligned} |\alpha^1| + \cdots + |\alpha^{N+1}| &\leq \sum_{\kappa=1}^{\kappa_0} \kappa(l(\kappa) - l(\kappa - 1)) \\ &= \sum_{\kappa=1}^{\kappa_0} \kappa(l(\kappa) - l(\kappa - 1)) \\ &= \kappa_0 l(\kappa_0) - (l(0) + \cdots + l(\kappa_0 - 1)) \\ &\leq \kappa_0(N + 1) - (1 + \cdots + \kappa_0) \\ &= \frac{N(N + 1) - (N - \kappa_0)^2 - (N - \kappa_0)}{2} \leq \frac{N(N + 1)}{2}. \end{aligned}$$ □

From now on, we fix α^j, $1 \leq j \leq N + 1$, obtained from Theorem 6.2.1, so that $W_{\alpha^1 \cdots \alpha^{N+1}}(f) \not\equiv 0$, $|\alpha^j| \leq N$ for $1 \leq j \leq N + 1$, and $|\alpha^1| + \cdots + |\alpha^{N+1}| \leq N(N + 1)/2$. Let $H_j = \{a_{j0}w_0 + \cdots + a_{jN}w_N = 0\}, 1 \leq j \leq q$, be hyperplanes in $\mathbb{P}^N(\mathbb{C})$ located in general position. Write $F_j := a_{j0}f_0 + \cdots + a_{jN}f_N$ for $1 \leq j \leq q$.

Theorem 6.2.5. *In the above situation, set $l_0 = |\alpha^1| + \cdots + |\alpha^{N+1}|$ and take t and p' such that $0 < tl_0 < p' < 1$. Then, for $0 < r_0 < R_0$, there exists a positive constant K such that for $r_0 < r < R < R_0$,*

$$\int_{S(r)} \left| z^{\alpha^1 + \cdots + \alpha^{N+1}} \frac{W_{\alpha^1 \cdots \alpha^{N+1}}(f)}{F_1 \cdots F_q} \right|^t \|f\|^{t(q-N-1)} \sigma_n \leq K \left(\frac{R^{2n-1}}{R - r} T_f(R) \right)^{p'}.$$

Proof. Since $H_1, \dots, H_q$ are in general position, similarly to Theorem 4.2.2, we have

$$\left| \frac{W_{\alpha^1 \cdots \alpha^{N+1}}(f)}{F_1 \cdots F_q} \right| \|f\|^{q-N-1} \leq K \left(\sum_{1 \leq j_1 < \cdots < j_{N+1} \leq q} \left| \frac{W_{\alpha^1 \cdots \alpha^{N+1}}(F_{j_1}, \dots, F_{j_{N+1}})}{F_{j_1} \cdots F_{j_{N+1}}} \right| \right)$$

for some constant $K > 0$. Hence we only need to estimate

$$I := \int_{S(r)} \left| z^{\alpha^1+\cdots+\alpha^{N+1}} \frac{W_{\alpha^1\cdots\alpha^{N+1}}(F_{j_1},\ldots,F_{j_{N+1}})}{F_{j_1}\cdots F_{j_{N+1}}} \right|^t \sigma_n.$$

Let

$$\chi := \frac{W_{\alpha^1\cdots\alpha^{N+1}}(F_{j_1},\ldots,F_{j_{N+1}})}{F_{j_1}\cdots F_{j_{N+1}}}$$

and set $\varphi_i := F_i/F_1, 1 \le i \le q$. Again, by the properties of the Wronskian we can write χ as

$$\chi = \frac{W_{\alpha^1\cdots\alpha^{N+1}}(\varphi_1,\varphi_2,\ldots,\varphi_{N+1})}{\varphi_1\cdots\varphi_{N+1}} = \sum_{\sigma_{i_1\cdots i_{N+1}}} \operatorname{sgn}(\sigma_{i_1\cdots i_{N+1}}) \frac{D^{\alpha^1}\varphi_{i_1}}{\varphi_{i_1}} \cdots \frac{D^{\alpha^{N+1}}\varphi_{i_{N+1}}}{\varphi_{i_{N+1}}}.$$

Therefore the above integrand can be estimated from above by a positive constant multiple of the sum of some functions of the type

$$\psi_{i_1\ldots i_{N+1}} := \left| z^{\alpha^1+\cdots+\alpha^{N+1}} \frac{D^{\alpha^1}\varphi_{i_1}}{\varphi_{i_1}} \cdots \frac{D^{\alpha^{N+1}}\varphi_{i_{N+1}}}{\varphi_{i_{N+1}}} \right|^t.$$

Set $p_j = \frac{|\alpha^j|}{|\alpha^1|+\cdots+|\alpha^{N+1}|}$ for $1 \le j \le N+1$. By the Hölder inequality we obtain

$$\int_{S(r)} \psi_{i_1\ldots i_{N+1}} \sigma_n \le \prod_{j=1}^{N+1} \left[\int_{S(r)} \left| z^{\alpha^j} \frac{D^{\alpha_j}\varphi_{i_j}}{\varphi_{i_j}} \right|^{t/p_j} \sigma_n \right]^{p_j}.$$

Since $(t/p_j)|\alpha^j| = (|\alpha_1|+\cdots+|\alpha_{N+1}|)t < p' < 1$ for $1 \le j \le N+1$, we can apply Corollary 6.1.4 to get

$$\int_{S(r)} \psi_{i_1\ldots i_{N+1}} \sigma_n \le K \prod_{j=1}^{N+1} \left(\frac{R^{2n-1}}{R-r} T_{\varphi_{i_j}}(R) \right)^{p' p_j}.$$

On the other hand, we have

$$T_{\varphi_j}(r) \le T_f(r) + K$$

for every $j = 1,\ldots,N+1$. Therefore we conclude

$$I \le K \left(\frac{R^{2n-1}}{R-r} T_f(R) \right)^{p'}.$$

This completes the proof of the theorem. □

Theorem 6.2.6 (Second main theorem). *Let $f : B(R_0) \subset \mathbb{C}^n \longrightarrow \mathbb{P}^N(\mathbb{C})$, $0 < R_0 \le \infty$, be a linearly nondegenerate meromorphic map, and let $H_1, \dots, H_q$ be hyperplanes in $\mathbb{P}^N(\mathbb{C})$ located in general position. Then, for all $0 < r < R_0$,*

$$(q - N - 1)T_f(r) \le \sum_{j=1}^{q} N_f^{(N)}(r, H_j) + S(r),$$

where $S(r)$ is evaluated as follows:

(1) *In the case $R_0 < \infty$,*

$$S(r) \le K\left(\log^+ \frac{1}{R_0 - r} + \log^+ T_f(r)\right)$$

for every $r \in [0, R_0)$ excluding a set E such that $\int_E \frac{1}{R_0 - t} dt < \infty$.

(2) *In the case $R_0 = \infty$,*

$$S(r) \le K(\log^+ T_f(r) + \log r)$$

for every $r \in [0, \infty)$ excluding a set E' such that $\int_{E'} dt < \infty$.

Proof. Let $L_j, 1 \le j \le q$, be the linear forms associated with the hyperplanes $H_1, \dots, H_q$. Theorem 6.2.5 implies that

$$\int_{S(r)} \left| z^{\alpha^1 + \dots + \alpha^{N+1}} \frac{W_{\alpha^1 \dots \alpha^{N+1}}(f)}{L_1(f) \cdots L_q(f)} \right|^t \|f\|^{t(q-N-1)} \sigma_n \le K\left(\frac{R^{2n-1}}{R - r} T_f(R)\right)^{p'} \tag{6.9}$$

for $r_0 < r < R < R_0$. By the concavity of the logarithm this inequality implies that

$$\begin{aligned} t \int_{S(r)} \log|z^{\alpha^1 + \dots + \alpha^{N+1}}| \sigma_n &+ t \int_{S(r)} \log\left| \frac{W_{\alpha^1 \dots \alpha^{N+1}}(f)}{L_1(f) \cdots L_q(f)} \right| \sigma_n \\ &+ t(q - N - 1) \int_{S(r)} \log\|f\| \sigma_n \\ &\le p'\left(\log^+ \frac{R}{R - r} + \log^+ T_f(R)\right) + O(1). \end{aligned} \tag{6.10}$$

However, similarly to Lemma 4.2.4, by the Jensen formula we can prove that

$$-\sum_{j=1}^{q} N_f^{(N)}(r, H_j) \le \int_{S(r)} \log\left| \frac{W_{\alpha^1 \dots \alpha^{N+1}}(f)}{L_1(f) \cdots L_q(f)} \right| \sigma_n + O(1).$$

Therefore

$$(q - N - 1)T_f(r) \leq \sum_{j=1}^{q} N_f^{(N)}(r, H_j) + C\left(\log r + \log^+ \frac{R}{R - r} + \log^+ T_f(R)\right) \tag{6.11}$$

for some constant $C > 0$. Since $T_f(r)$ is continuous and increasing and we may assume that $T_f(r) \geq 1$, we can apply Lemma 4.1.7 to show

$$T_f\left(r + \frac{R_0 - r}{eT_f(r)}, r_0\right) \leq 2T_f(r)$$

outside a set E of r such that $\int_E 1/(R_0 - r)dr < \infty$ in the case $R_0 < \infty$ and

$$T_f\left(r + \frac{1}{T_f(r)}\right) \leq 2T_f(r)$$

outside a set E' of r such that $\int_{E'} dr < \infty$ in the case $R_0 = \infty$. Substituting $R = r + \frac{R_0 - r}{eT_f(r)}$ if $R_0 < \infty$ and $R = r + 1/T_f(r)$ if $R_0 = \infty$ into (6.11) proves the theorem. □

Corollary 6.2.7. *In the situation of Theorem* 6.2.6, *if*

$$\text{(i)} \quad \limsup_{r \to R_0} \frac{T_f(r)}{\log(1/(R_0 - r))} = \infty$$

or

$$\text{(ii)} \quad R_0 = \infty,$$

then

$$\sum_j \delta_N^{f,*}(H_j) \leq N + 1, \tag{6.12}$$

where $\delta_N^{f,*}$ *is the classical Nevanlinna defect defined as*

$$\delta_N^{f,*}(H_j) = 1 - \limsup_{r \to R_0} \frac{N_f^{(N)}(r, H_j)}{T_f(r)}.$$

6.3 The nonintegrated defect relation

Let M be an n-dimensional complex Kähler manifold. Let f be a meromorphic map of M into $\mathbb{P}^N(\mathbb{C})$, let μ_0 be a positive integer, and let D be a hypersurface in $\mathbb{P}^N(\mathbb{C})$ of degree d such that $f(M) \not\subset D$. We denote the intersection multiplicity of the image of f and D at $f(p)$ by $\nu^f(D)(p)$ and the pullback of the normalized Fubini–Study metric form Ω on $\mathbb{P}^N(\mathbb{C})$ by Ω_f. The *nonintegrated defect of* f *with respect to* D *cut by* μ_0 is defined as

$$\delta^f_{\mu_0}(D) := 1 - \inf\{\eta \geq 0 : \eta \text{ satisfies condition } (\star)\}.$$

Here condition $(\star)$ means that there exists a bounded nonnegative continuous function h on M with zeros of order not less than $\min(\nu^f(D), \mu_0)$ such that

$$d\eta\Omega_f + \frac{\sqrt{-1}}{2\pi}\partial\bar{\partial}\log h^2 \geq [\min(\nu^f(D), \mu_0)],$$

where d is the degree of D, and by $[\nu]$ we mean the (1, 1)-current associated with a divisor ν. Note that condition $(\star)$ also means that for each holomorphic function $\phi(\not\equiv 0)$ on an open subset U of M with $\nu_\phi = \min(\nu^f(D), \mu_0)$ outside an analytic set of codimension ≥ 2, the function $u := \log(h^2\|f\|^{2d\eta}/|\phi|^2)$ is continuous and plurisubharmonic on U, where $\|f\|^2 = |f_0|^2 + \cdots + |f_N|^2$, and $f = [f_0 : \cdots : f_N]$ is a (local) reduced representation of f. Similarly to the classical Nevanlinna defect, we have the following properties.

- $0 \leq \delta^f_{\mu_0}(D) \leq 1$. To see that $\delta^f_{\mu_0}(D) \geq 0$, take $\eta = 1$ and $h = |Q(f)|/\|f\|^d$, where Q is the homogeneous polynomial defining D;
- If $f(M) \cap D = \emptyset$, then by taking $\eta = 0$ and $h = 1$ we have that $\delta^f_{\mu_0}(D) = 1$;
- If $\nu^f(D)(p) \geq \mu$ for all $p \in f^{-1}(D)$ with some positive integer $\mu \geq \mu_0$, then $\delta^f_{\mu_0}(D) \geq 1 - \mu_0/\mu$ by taking $\eta = \mu_0/\mu$ and $h = |Q(f)|/\|f\|^{\mu_0 d/\mu}$.

Let $f : B(R_0) \subset \mathbb{C}^n \longrightarrow \mathbb{P}^N(\mathbb{C})$, $0 < R_0 \leq \infty$, be a meromorphic map. Note that the classical Nevanlinna defect $\delta^{f,\star}_{\mu_0}(D)$ of f with respect to D of degree d cut by μ_0 is defined as

$$\delta^{f,\star}_{\mu_0}(D) = 1 - \limsup_{r \to R_0} \frac{N^{(\mu_0)}_f(r, D)}{dT_f(r)}.$$

The relationship between the nonintegrated defect and the classical Nevanlinna defect is given as follows.

Proposition 6.3.1. *Let $f : B(R_0) \subset \mathbb{C}^n \longrightarrow \mathbb{P}^N(\mathbb{C})$, $0 < R_0 \leq \infty$, be a meromorphic map. If $\lim_{r \to R_0} T_f(r) = \infty$, then*

$$0 \leq \delta^f_{\mu_0}(D) \leq \delta^{f,\star}_{\mu_0}(D) \leq 1,$$

where $\delta^\star$ is the classical Nevanlinna defect.

Proof. Take η satisfying condition $(\star)$ in the definition of $\delta^f_{\mu_0}(D)$. Then the function

$$v := d\eta \log\|f\| + \log h - \log|\varphi|$$

is plurisubharmonic, where h is bounded, and φ is holomorphic on $B(R_0)$ with $\nu_\varphi = \min(\nu^f(D), \mu_0)$ outside an analytic set of codimension ≥ 2. Therefore

$$\begin{aligned}0 &\le \int_{S(r)} v\sigma_n - \int_{S(r_0)} v\sigma_n \\ &= d\eta \int_{S(r)} \log\|f\|\sigma_n + \int_{S(r)} \log h\sigma_n - \int_{S(r)} \log|\varphi|\sigma_n + K \\ &\le d\eta T_f(r) - N_f^{(\mu_0)}(r, D) + K,\end{aligned}$$

where K is a constant, because h is bounded from above. This implies that

$$\frac{N_f^{(\mu_0)}(r, D)}{dT_f(r)} \le \eta + \frac{K}{T_f(r)}.$$

As $r \to R_0$, we obtain $\delta_{\mu_0}^{f,\star}(D) \ge 1 - \eta$. Hence $\delta_{\mu_0}^{f,\star}(D) \ge \delta_{\mu_0}^{f}(D)$. □

Let M be an n-dimensional complex Kähler manifold. Let $\omega = \frac{\sqrt{-1}}{2} \sum_{ij} h_{i\bar{j}} dz_i \wedge d\bar{z}_j$ be the Kähler form of M. We define

$$\mathrm{Ric}(\omega) = dd^c \log(\det(h_{i\bar{j}})),$$

where $d = \partial + \bar{\partial}$ and $d^c = \frac{\sqrt{-1}}{4\pi}(\bar{\partial} - \partial)$. Let f be a meromorphic map of M into $\mathbb{P}^N(\mathbb{C})$. We assume the following growth condition for f: there exists a nonzero bounded continuous real-valued function h on M such that

$$\rho\Omega_f + \frac{\sqrt{-1}}{2\pi}\partial\bar{\partial}\log h^2 \ge \mathrm{Ric}(\omega) \tag{6.13}$$

for some nonnegative constant ρ.

Theorem 6.3.2 ([26]). *Let M be an n-dimensional complete Kähler manifold, and let $f : M \to \mathbb{P}^N(\mathbb{C})$ be a linearly nondegenerate meromorphic map. Assume that there exists a nonzero bounded continuous real-valued function h on M such that*

$$\rho\Omega_f + \frac{\sqrt{-1}}{2\pi}\partial\bar{\partial}\log h^2 \ge Ric(\omega) \tag{6.14}$$

for some nonnegative constant ρ. Assume that the universal covering of M is biholomorphic to a ball $B(R_0) \subset \mathbb{C}^n$ with $0 < R_0 \le \infty$. Let $H_1, \dots, H_q$ be hyperplanes in $\mathbb{P}^N(\mathbb{C})$ located in general position. Then

$$\sum_{j=1}^{q} \delta_N^f(H_j) \le N + 1 + \rho N(N+1).$$

Proof. Since the universal covering of M is $B(R_0) \subset \mathbb{C}^n$, by lifting f to the covering, we may assume that $M = B(R_0) \subset \mathbb{C}^n$. In the case where $R_0 = \infty$ and the case where

$$\lim_{r\longrightarrow R_0} \sup \frac{T_f(r)}{\log 1/(R_0 - r)} = \infty,$$

from Proposition 6.3.1 and Corollary 6.2.7 we have

$$\sum_{j=1}^{q} \delta_N^f(H_j) \leq \sum_{j=1}^{q} \delta_N^{f,*}(H_j) \leq N + 1.$$

Hence this case is proved. Now we consider the remaining case where $R_0 < \infty$. Without loss of generality, we assume that $R_0 = 1$ under the condition that

$$\lim_{r\longrightarrow 1} \sup \frac{T_f(r)}{\log 1/(1 - r)} < \infty.$$

We assume that

$$\sum_{j=1}^{q} \delta_N^f(H_j) > N + 1 + \rho N(N + 1).$$

By the definition of nonintegrated defect, there exist constants $\eta_j \geq 0$ and continuous plurisubharmonic functions $u_j(\not\equiv -\infty)$ such that

$$\frac{\rho N(N + 1)}{(1 - \eta_1) + \cdots + (1 - \eta_q) - N - 1} < 1$$

and

$$e^{u_j}|\varphi_j| \leq \|f\|^{\eta_j}$$

for $j = 1, \ldots, q$, and $u_j - \log|\varphi_j|$ is plurisubharmonic, where φ_j is a nonzero holomorphic function with $\nu_{\varphi_j}^0 = \min(\nu^f(H_j), N)$. Since f is linearly nondegenerate, by Theorem 6.2.1 there exist $\alpha^j = (\alpha_{j1}, \ldots, \alpha_{jn})$ with integer $\alpha_{ji} \geq 0$ such that $|\alpha^j| \leq N$ for $1 \leq j \leq N + 1$, $|\alpha^1| + \cdots + |\alpha^{N+1}| \leq N(N+1)/2$, and $W_{\alpha^1\ldots\alpha^{N+1}}(f) \not\equiv 0$. Denote $H_j = \{a_{j0}w_0 + \cdots + a_{jN}w_N = 0\}$ and $F_j = a_{j0}f_0 + \cdots + a_{jN}f_N$ for $1 \leq j \leq q$. Let

$$v := \log\left|z^{\alpha^1+\cdots+\alpha^{N+1}} \frac{W_{\alpha^1\ldots\alpha^{N+1}}(f)}{F_1\cdots F_q}\right| + \sum_{j=1}^{q} u_j. \tag{6.15}$$

We now show that v is plurisubharmonic on $M = B(1)$. To do so, we set

$$\psi = \frac{W_{\alpha^1\ldots\alpha^{N+1}}(f)}{F_1\cdots F_q}.$$

We claim that

$$\nu_\psi^\infty \le \sum_{j=1}^{q} \min\{\nu_{F_j}^0, N\} \tag{6.16}$$

outside an analytic set of codimension at least two. To prove this, let I_f be the indeterminacy set of f, and take $a \in B(1)\backslash I_f$. We first show the following claim: *Let h be a holomorphic function around a. Assume that $D^\alpha h \not\equiv 0$ around a. Then*

$$\nu^\infty_{\frac{D^\alpha h}{h}}(a) = \min\{\nu_h^0(a), |\alpha|\}. \tag{6.17}$$

To see this, take a system of holomorphic local coordinates $z = (z_1, \dots, z_n)$ in a neighborhood of a such that $z(a) = 0$ and h can be written as $h = z_1^{\nu_h^0(a)}\tilde{h}$ with $\tilde{h}$ having no zero in a neighborhood of a. From this representation of h we can conclude that $\nu_{D^\alpha h}^0(a) = \max\{0, \nu_h^0(a) - |\alpha|\}$. Thus (6.17) holds.

Now for each $a \in B(1)\backslash I_f$, without loss of generality, we may assume that F_j vanishes at a for $1 \le j \le q_1$ and F_j does not vanish at a for $j > q_1$. By the assumption that the F_j are in general position we know that $q_1 \le N$. Write

$$\gamma := \frac{W_{\alpha^1 \cdots \alpha^{N+1}}(F_1, \dots, F_{N+1})}{F_1 \cdots F_{N+1}}.$$

Then $\nu_\psi^\infty = \nu_\gamma^\infty$ since F_j does not vanish at a for $j > N$ and

$$W_{\alpha^1 \cdots \alpha^{N+1}}(F_1, \dots, F_{N+1}) = cW_{\alpha^1 \cdots \alpha^{N+1}}(f).$$

On the other hand, we can write

$$\begin{aligned}\gamma &= \det(D^{\alpha_i} F_j / F_j : 1 \le i, j \le N+1)\\ &= \sum \operatorname{sgn}(i_1 \cdots i_{N+1}) \frac{D^{\alpha_1} F_1}{F_1} \cdots \frac{D^{\alpha_{N+1}} F_{N+1}}{F_{N+1}}.\end{aligned}$$

By (6.17) and the fact that $|\alpha_j| \le N$ for $1 \le j \le N+1$ we get

$$\nu_\gamma^\infty(a) \le \sum_{j=1}^{N+1} \min\{\nu_{F_j}^0(a), N\} \le \sum_{j=1}^{q} \min\{\nu_{F_j}^0(a), N\}.$$

This proves the claim. Since $u_j - \log|\varphi_j|$ is plurisubharmonic and $\nu_{\varphi_j}^0 = \min(\nu^f(H_j), N)$, by the definition of v and (6.16), we see that v is plurisubharmonic on $M = B(1)$.

By the growth condition of f (see (6.14)) there exists a continuous plurisubharmonic function $w \not\equiv -\infty$ on $B(1)$ such that

$$e^w dV \le \|f\|^{2\rho} v_n^n. \tag{6.18}$$

Set

$$t = \frac{2\rho}{q - N - 1 - (\eta_1 + \cdots + \eta_q)} \tag{6.19}$$

and

$$\chi := z^{\alpha^1+\cdots+\alpha^{N+1}} \frac{W_{\alpha^1\ldots\alpha^{N+1}}(f)}{F_1 \cdots F_q}.$$

Define

$$u := w + tv.$$

Then u is plurisubharmonic and thus subharmonic on the Kähler manifold M. We then have

$$e^u dV \leq e^{w+tv} dV \leq e^{tv} \|f\|^{2\rho} v_n^n \leq |\chi|^t \|f\|^{t(q-n-1)} v_n^n.$$

Therefore

$$\int_{B(1)} e^u dV = \int_{B(1)} e^u v_n^n \leq 2n \int_0^1 r^{2n-1} \Big(\int_{S(r)} |\chi|^t \|f\|^{t(q-n-1)} \sigma_n \Big) dr,$$

where we used the identity $v_n^n = (dd^c|z|^2)^n = 2n|z|^{2n-1}\sigma_n \wedge d|z|$. Note that $t \cdot \frac{N(N+1)}{2} < 1$, so we take p' such that $t \cdot \frac{N(N+1)}{2} < p' < 1$. Then

$$0 < (|\alpha_1| + \cdots + |\alpha_{n+1}|)t \leq t \cdot \frac{N(N+1)}{2} < p' < 1.$$

Therefore by Theorem 6.2.5

$$\int_{S(r)} |\chi|^t \|f\|^{t(q-n-1)} \sigma_n \leq K \Big(\frac{R^{2n-1}}{R - r} T_f(R) \Big)^{p'}$$

for $r_0 < r < R < 1$. According to Lemma 4.1.7, choose $R := r + (1 - r)/eT_f(r)$. Then

$$T_f(R) \leq 2T_f(r)$$

outside a set such that $\int_E 1/(1 - r) dr < \infty$. Thus

$$\int_{S(r)} |\chi|^t \|f\|^{t(q-n-1)} \sigma_n \leq K(T_f(R))^{p'} \leq K \frac{1}{(1 - r)^{p'}} \Big(\log \frac{1}{1 - r} \Big)^{p'}$$

for all $r \in [r_0, 1)\backslash E$. Slightly varying the constant K, we may assume that this inequality holds for all $r \in [r_0, 1)$ because of Proposition 4.1.8. From these facts we conclude that

$$\int_{B(1)} e^u dV \leq K \int_0^1 \frac{t^{2n-1}}{(1-t)^{p'}} \left(\log \frac{1}{1-t}\right)^{p'} dt < \infty.$$

On the other hand, by Theorem 3.3.9 (see Yau [90] or Karp [45]) we necessarily have

$$\int_{B(1)} e^u dV = \infty,$$

because B(1) has infinite volume with respect to the given complete Kähler metric (cf. [45], Theorem B). This is a contradiction. So the proof of the theorem is completed. □

The rest of this chapter is to extend Theorem 6.3.2 to the case where $f(M)$ intersects with hypersurfaces instead of hyperplanes; namely, we prove the following main theorem.

Theorem 6.3.3 ([77]). *Let M be an n-dimensional complete Kähler manifold, and let $f : M \to \mathbb{P}^N(\mathbb{C})$ be an algebraically nondegenerate meromorphic map (i. e., its image is not contained in any proper subvariety of $\mathbb{P}^N(\mathbb{C})$). Assume that the universal covering of M is biholomorphic to a ball $B(R_0)$ in $\mathbb{C}^n$ with $0 < R_0 \leq \infty$. Let $D_1, \ldots, D_q$ be hypersurfaces of degree d_j in $\mathbb{P}^N(\mathbb{C})$ located in general position. Let $d = \text{l. c. m.}\{d_1, \ldots, d_q\}$ (the least common multiple of $\{d_1, \ldots, d_q\}$). Assume that there exists a nonzero bounded continuous real-valued function h on M such that $\rho\Omega_f + \frac{\sqrt{-1}}{2\pi}\partial\bar{\partial}\log h^2 \geq \text{Ric}(\omega)$ for some nonnegative constant ρ, where ω is the Kähler form of M. Then, for every $\epsilon > 0$,*

$$\sum_{j=1}^{q} \delta_{l-1}^f(D_j) \leq N + 1 + \epsilon + \frac{\rho l(l-1)}{d},$$

where $l \leq 2^{N^2+4N} e^N d^{2N} (NI(\epsilon^{-1}))^N$, and $I(x) := \min\{k \in \mathbb{N} : k > x\}$ for a positive real number x.

Proof. Since the universal covering of M is the unit ball $B(R_0) \subset \mathbb{C}^n$, by lifting f to the covering we may assume that $M = B(R_0) \subset \mathbb{C}^m$. Let $D_1, \ldots, D_q$ be hypersurfaces in $\mathbb{P}^N(\mathbb{C})$ of degrees $d_1, \ldots, d_q$ located in general position. Let $Q_j, 1 \leq j \leq q$, be the homogeneous polynomials defining D_j. Replacing Q_j by Q_j^{d/d_j} if necessary, where d is the l. c. m. (the least common multiple) of d_j, we can assume that $Q_1, \ldots, Q_q$ have the same degree d. For $m \in \mathbb{N}$, let V_m be the space of homogeneous polynomials of $N+1$ variables of degree m and fix an arbitrary basis $\phi_1, \ldots, \phi_l$ of V_m, where $l = \dim V_m$. Set $F = [\phi_1(f) : \cdots : \phi_l(f)]$. Then $F : B(R_0) \to \mathbb{P}^{l-1}(\mathbb{C})$ is linearly nondegenerate. By Theorem 6.2.1 there exist $\alpha^j = (\alpha_{j1}, \ldots, \alpha_{jl})$ with integer $\alpha_{ji} \geq 0$ such that $|\alpha^j| \leq l-1$ for $1 \leq j \leq l$, $|\alpha^1| + \cdots + |\alpha^l| \leq l(l-1)/2$, and the generalized Wronskian $W_{\alpha^1 \ldots \alpha^l}(F) \not\equiv 0$. Similarly to (4.44), for $z \in B(R_0)$, there are linear forms $L_1, \ldots, L_u$ of l-variables such that

$$\frac{\|f(z)\|^{dq\triangle}}{|Q_1^{\triangle}(f(z))\cdots Q_q^{\triangle}(f(z))|} \le K'\frac{\|F(z)\|^l}{\prod_{t=1}^l |L_{t,z}(F(z))|} \le K\frac{\|f(z)\|^{lm}}{\prod_{t=1}^l |L_{t,z}(F(z))|},$$

where $\triangle$ is defined in (4.42), $K > 0$ is a constant, and $\{L_{1,z},\dots,L_{l,z}\} \subset \{L_1,\dots,L_u\}$ are linearly independent. Hence

$$\frac{\|f(z)\|^{dq\triangle-lm}|W_{\alpha^1\dots\alpha^l}(F)(z)|}{|Q_1^{\triangle}(f(z))\cdots Q_q^{\triangle}(f(z))|} \le K\frac{|W_{\alpha^1\dots\alpha^l}(F)(z)|}{|L_1(F(z))\cdots L_l(F(z))|}, \tag{6.20}$$

where we just write $L_{j,z}$ as L_j for simplicity.

We first deal with the case $R_0 < \infty$. We assume that $R_0 = 1$. So we let $f : B(1) \longrightarrow \mathbb{P}^N(\mathbb{C})$ be an algebraically nondegenerate meromorphic map. We break the proof into the following two cases: the case

$$\limsup_{r\longrightarrow 1} \frac{T_f(r)}{\log 1/(1-r)} < \infty$$

and the case

$$\limsup_{r\longrightarrow 1} \frac{T_f(r)}{\log 1/(1-r)} = \infty.$$

We first deal with the case where

$$\limsup_{r\longrightarrow 1} \frac{T_f(r)}{\log 1/(1-r)} < \infty.$$

We prove Theorem 6.3.3 by contradiction.

We assume that

$$\rho\Omega_f + \frac{\sqrt{-1}}{2\pi}\partial\bar\partial \log h^2 \ge \mathrm{Ric}(\omega) \tag{6.21}$$

and

$$\sum_{j=1}^q \delta_{l-1}^f(D_j) > (n+1) + \epsilon + \frac{\rho l(l-1)}{d}. \tag{6.22}$$

Then by the earlier discussion there exist constants $\eta_j \ge 0$ and continuous plurisubharmonic functions $\tilde u_j(\not\equiv -\infty)$ such that $e^{\tilde u_j}|\varphi_j| \le \|f\|^{d\eta_j}$ for $j = 1,\dots,q$ and

$$q - \sum_{j=1}^q \eta_j > n+1+\epsilon+\frac{\rho l(l-1)}{d}, \tag{6.23}$$

where φ_j is a nonzero holomorphic function with $\nu^0_{\varphi_j} = \min(\nu^f(D_j), l-1)$. Let $u_j = \tilde u_j + \log|\varphi_j|$. Then $u_j(\not\equiv -\infty), 1 \le j \le q$, are continuous plurisubharmonic functions,

$$e^{u_j} \le \|f\|^{d\eta_j}, \tag{6.24}$$

and $u_j - \log|\varphi_j|$ is plurisubharmonic, where φ_j is a nonzero holomorphic function with $\nu^0_{\varphi_j} = \min(\nu^f(D_j), l-1)$. Let

$$v := \log\left|z^{\alpha^1+\cdots+\alpha^l}\frac{W_{\alpha^1\ldots\alpha^l}(F)}{Q_1^{\triangle}(f)\cdots Q_q^{\triangle}(f)}\right| + \triangle\sum_{j=1}^{q} u_j. \tag{6.25}$$

We now show that v is plurisubharmonic on $M = B(1)$. To do so, set

$$\psi = \frac{W_{\alpha^1\ldots\alpha^l}(F)}{Q_1^{\triangle}(f)\cdots Q_q^{\triangle}(f)}.$$

We claim that

$$\nu_\psi^\infty \le \sum_{j=1}^{q}\triangle\min\{\nu^0_{Q_j(f)}, l-1\} \tag{6.26}$$

outside an analytic set of codimension at least two. Indeed, let I_F be the indeterminacy set of F, and take $a \in B(1)\backslash I_F$. For each $a \in B(1)\backslash I_F$, without loss of generality, we may assume that $Q_j(f)$ vanishes at a for $1 \le j \le q_1$ and does not vanish at a for $j > q_1$. By the assumption that the Q_j are in general position we know that $q_1 \le n$. For $\{Q_1,\ldots,Q_n\} \subset \{Q_1,\ldots,Q_q\}$, consider the filtration $V_m = W_{\mathbf{0}} \supset \cdots W_{\mathbf{i}} \supset W_{\mathbf{i}'} \supset \cdots$ defined in (4.41) associated with $\{Q_1,\ldots,Q_n\}$, and take a basis $\psi_1,\ldots,\psi_l$ of V_m according to this filtration. Then there are linearly independent linear forms $L_1,\ldots,L_l$ such that $\psi_t(f) = L_t(F), 1 \le t \le l$. Denote by $W := W_{\alpha^1\ldots\alpha^l}(F)$ the generalized Wronskian of F. By the basic properties of generalized Wronskian

$$W = W_{\alpha^1\ldots\alpha^l}(F) = CW_{\alpha^1\ldots\alpha^l}(L_1(F),\ldots,L_l(F)) = CW_{\alpha^1\ldots\alpha^l}(\psi_1(f),\cdots,\psi_l(f)),$$

where C is some constant. Let ψ be an element of the basis $\{\psi_1,\ldots,\psi_l\}$. As we discussed earlier, we may write $\psi = Q_1^{i_1}\cdots Q_n^{i_N}\eta$ with $\eta \in V_{m-d\sigma(i)}$. Therefore

$$\psi(f) = (Q_1(f))^{i_1}\cdots(Q_N(f))^{i_N}\eta(f).$$

Note that there are $\triangle_{\mathbf{i}}$ such ψ is our basis, where $\triangle_{\mathbf{i}} = \dim(W_{\mathbf{i}}/W_{\mathbf{i}'})$ with $\mathbf{i}' > \mathbf{i}$ are consecutive n-tuples with $W_{\mathbf{i}'} \subset W_{\mathbf{i}}$. Assume that $\nu^0_{Q_j(f)}(a) \ge l-1$ for $1 \le j \le q_0$ and $\nu^0_{Q_j(f)}(a) < l-1$ for $q_0 < j \le q_1$. Since, by the above, $W = C\det(D^{\alpha^i}(\psi_j(f)))_{1\le i,j\le l}$, using (6.17) and noticing that there are $\triangle_{\mathbf{i}}$ such ψ is our basis and that $|\alpha^j| \le l-1$ for $1 \le j \le l$, we have

$$\nu^0_W(a) \ge \sum_{\mathbf{i}}\left(\sum_{j=1}^{q_0} i_j(\nu^0_{Q_j}(a) - (l-1))\right)\triangle_{\mathbf{i}}$$

$$= \sum_{j=1}^{q_0}\Big(\sum_{\mathbf{i}} i_j \triangle_{\mathbf{i}}\Big)(v^0_{Q_j}(a) - (l-1)) = \triangle \sum_{j=1}^{q_0}(v^0_{Q_j}(a) - (l-1)).$$

On the other hand,

$$\sum_{j=1}^{q} v^0_{Q_j(f)}(a) = \sum_{j=1}^{N} v^0_{Q_j(f)}(a) = \sum_{j=1}^{q_0} v^0_{Q_j(f)}(a) + \sum_{j=q_0}^{q_1} v^0_{Q_j(f)}(a).$$

Hence $v^{\infty}_{\psi}(a) \le \sum_{j=0}^{q} \triangle \min\{v^0_{Q_j(f)}(a), l-1\}$. This proves (6.26). By the definition of v (see (6.25)), using (6.26) and the fact that $u_j - \log|\varphi_j|$ is plurisubharmonic and $v^0_{\varphi_j} = \min(v^f(D_j), l-1)$, we see that v is plurisubharmonic on $M = B(1)$.

We now continue our proof. By the growth condition of f (see (6.21)) there exists a continuous plurisubharmonic function $w \not\equiv -\infty$ on $B(1)$ such that

$$e^{w} dV \le \|f\|^{2\rho} v_n^n. \tag{6.27}$$

Set

$$t = \frac{2\rho}{qd\triangle - lm - \triangle d(\eta_1 + \cdots + \eta_q)} \tag{6.28}$$

and

$$\chi := z^{\alpha^1+\cdots+\alpha^l} \frac{W_{\alpha^1\cdots\alpha^l}(F)}{Q_1^{\triangle}(f)\cdots Q_q^{\triangle}(f)}.$$

Define

$$u := w + tv.$$

Then u is plurisubharmonic and thus subharmonic on the Kähler manifold M.

By Theorem 3.3.9 (or Karp [45]) we necessarily have

$$\int_{B(1)} e^{u} dV = \infty,$$

because $B(1)$ has infinite volume with respect to the given complete Kähler metric (cf. [45], Theorem B). Now by (6.24), (6.27), and (6.28)

$$\begin{aligned} e^{u} dV &= e^{w+tv} dV \le e^{tv}\|f\|^{2\rho} v_n^n \\ &= |\chi|^t \Big(\prod_{j=1}^{q} e^{t\triangle u_j}\Big)\|f\|^{2\rho} v_n^n \le |\chi|^t \Big(\prod_{j=1}^{q} \|f\|^{t\triangle d\eta_j}\Big)\|f\|^{2\rho} v_n^n \\ &= |\chi|^t \|f\|^{2\rho + td\triangle\sum_{j=1}^{q}\eta_j} v_n^n = |\chi|^t\|f\|^{t(dq\triangle - lm)} v_n^n. \end{aligned}$$

The contradiction will appear if we show that

$$\int_{B(1)} e^u dV < \infty.$$

By Lemma 4.2.13, $\frac{lm}{\triangle} \leq d(N+1) + \epsilon$. Thus $qd - \frac{lm}{\triangle} \geq d(q - (N+1+\epsilon))$. So, using (6.23), we have

$$dq \triangle - lm - \triangle \sum_{j=1}^{q} d\eta_j \geq d \triangle (q - (N+1+\epsilon)) - \triangle \sum_{j=1}^{q} d\eta_j > \triangle pl(l-1).$$

This implies that $tl(l-1)/2 < 1$. Since $|\alpha^1| + \cdots + |\alpha^l| \leq l(l-1)/2$, we can choose p' such that $t(|\alpha^1| + \cdots + |\alpha^l|) \leq tl(l-1)/2 < p' < 1$. By the identity

$$v_n^n = (dd^c|z|^2)^n = 2n|z|^{2n-1}\sigma_n \wedge d|z|$$

we have

$$\begin{aligned}\int_{B(1)} e^u dV &\leq \int_{B(1)} |\chi|^t \|f\|^{t(dq\triangle - lm)} v_n^n \\ &\leq 2n \int_0^1 r^{2n-1}\Big(\int_{S(r)} |\chi|^t \|f\|^{t(dq\triangle - lm)} \sigma_n\Big) dr \\ &= 2n \int_0^1 r^{2n-1}\Big(\int_{S(r)} \Big| z^{\alpha^1+\cdots+\alpha^l} \frac{W_{\alpha^1\cdots\alpha^l}(F)\|f\|^{(dq\triangle - lm)}}{Q_1^{\triangle}(f)\cdots Q_q^{\triangle}(f)} \sigma_n\Big|^t\Big) dr. \end{aligned} \tag{6.29}$$

On the other hand, by (6.20)

$$\frac{|W_{\alpha^1\cdots\alpha^l}(F)|\|f\|^{(dq\triangle - lm)}}{|Q_1^{\triangle}(f)\cdots Q_q^{\triangle}(f)|} \leq K \sum_{L_1,\ldots,L_l} \Big(\frac{|W_{\alpha^1\cdots\alpha^l}(F)|}{|L_1(F)\cdots L_l(F)|}\Big), \tag{6.30}$$

where the summation is taken over all possible linear forms $L_1, \ldots, L_l$, which is a finite summation. By Theorem 6.2.5, for all $L_1, \ldots, L_l$,

$$\int_{S(r)} \Big| z^{\alpha^1+\cdots+\alpha^l} \frac{W_{\alpha^1\cdots\alpha^l}(F)}{L_1(F)\cdots L_l(F)}\Big|^t \sigma_n \leq K\Big(\frac{R^{2n-1}}{R-r} T_F(R)\Big)^{p'}. \tag{6.31}$$

Combining (6.30) and (6.31) thus gives

$$\int_{S(r)} \Big| z^{\alpha^1+\cdots+\alpha^l} \frac{W_{\alpha^1\cdots\alpha^l}(F)}{Q_1^{\triangle}(f)\cdots Q_q^{\triangle}(f)}\Big|^t \|f\|^{t(dq\triangle - lm)} \sigma_n \leq K\Big(\frac{R^{2n-1}}{R-r} T_F(R)\Big)^{p'} \tag{6.32}$$

for $r_0 < r < R < 1$, where, as we noted, we use the letter K to denote a constant depending only on f and $D_1, \dots, D_q$, even when it should be replaced by a new constant. According to Lemma 4.1.7, if we choose $R := r + (1-r)/eT_F(r)$, then

$$T_F(R) \le 2T_F(r) \le 2dT_f(r)$$

outside a set E such that $\int_E 1/(1-r)dr < \infty$. If

$$\limsup_{r \to 1} \frac{T_f(r)}{\log 1/(1-r)} < \infty,$$

then (6.32) becomes

$$\int\limits_{S(r)} \left| z^{\alpha^1 + \dots + \alpha^l} \frac{W_{\alpha^1 \dots \alpha^l}(F)}{Q_1^{\triangle}(f) \cdots Q_q^{\triangle}(f)} \right|^t \|f\|^{t(dq\triangle - lm)} \sigma_n \le \frac{K}{(1-r)^{p'}} \left(\log \frac{1}{1-r} \right)^{p'} \tag{6.33}$$

for all $r \in [0,1)$ outside a set E such that $\int_E 1/(1-r)dr < \infty$. Slightly varying a constant K, we may assume that this inequality holds for all $r \in [0,1)$ because of Proposition 4.1.8. Therefore by (6.29) and (6.33) we have

$$\int\limits_{B(1)} e^u dV \le K \int\limits_0^1 \frac{r^{2n-1}}{(1-r)^{p'}} \left(\log \frac{1}{1-r} \right)^{p'} dr < \infty,$$

since $p' < 1$. This gives a contradiction and thus completes the proof for the first case.

We now deal with the case where either $R_0 = \infty$ or $R_0 < \infty$ with

$$\limsup_{r \to 1} \frac{T_f(r)}{\log 1/(R_0 - r)} = \infty.$$

Similarly to the proof of (6.32), using (6.20) and Theorem 6.2.5, we have

$$\int\limits_{S(r)} \left| z^{\alpha^1 + \dots + \alpha^l} \frac{W_{\alpha^1 \dots \alpha^l}(F)}{Q_1^{\triangle}(f) \cdots Q_q^{\triangle}(f)} \right|^t \|f\|^{t(dq\triangle - lm)} \sigma_n \le K \left(\frac{R^{2n-1}}{R-r} T_F(R) \right)^{p'} \tag{6.34}$$

for $r_0 < r < R < R_0$. Hence, by the concavity of the logarithm this inequality implies that

$$\begin{aligned} t \int\limits_{S(r)} \log |z^{\alpha^1 + \dots + \alpha^l}| \sigma_n &+ t \int\limits_{S(r)} \log \left| \frac{W_{\alpha^1 \dots \alpha^l}(F)}{Q_1^{\triangle}(f) \cdots Q_q^{\triangle}(f)} \right| \sigma_n \\ &+ t(dq \triangle - lm) \int\limits_{S(r)} \log \|f\| \sigma_n \\ &\le p' \left(\log^+ \frac{R^{2n-1}}{R-r} + \log^+ T_F(R) \right) + O(1) \end{aligned} \tag{6.35}$$

for $r_0 < R < R_0$. However, by the Jensen formula (see Corollary 4.1.3) and (6.26)

$$\int_{S(r)} \log\left|\frac{W_{\alpha^1\dots\alpha^l}(F)}{Q_1^{\triangle}(f)\cdots Q_q^{\triangle}(f)}\right|\sigma_n \le \triangle \sum_{j=1}^{q} N_f^{[l-1]}(r, D_j).$$

Also $T_F(R) \le mT_f(R) + O(1)$. Therefore (6.35) becomes

$$(dq\triangle - lm)T_f(r) \le \sum_{j=1}^{q} \triangle N_f^{[l-1]}(r, D_j)$$
$$+ K\left(\log r + \log^+ \frac{R}{R-r} + \log^+ T_f(R)\right) + O(1).$$

By Lemma 4.2.13 with $m = 2d(N+1)(Nd+N)(2^N - 1)(I(\epsilon^{-1}) + 1) + Nd$ for any $\epsilon > 0$ we have

$$\frac{lm}{\triangle} \le d(N+1) + \epsilon,$$

and, moreover, l satisfies $l \le 2^{N^2+4N} e^N d^{2N} (NI(\epsilon^{-1}))^N$. Hence

$$(q - N - 1 - \epsilon)T_f(r) \le \sum_{j=1}^{q} d^{-1} N_f^{[l-1]}(r, D_j)$$
$$+ K\left(\log r + \log^+ \frac{R}{R-r} + \log^+ T_f(R)\right).$$

In the case $R_0 = \infty$, we take $R = r + \frac{1}{T_f(r)}$. By Lemma 4.1.7 we still have $T_f(R) \le T_f(r) + 1$ for all r outside a set E of finite measure. Thus we have

$$(q - N - 1 - \epsilon)T_f(r) \le \sum_{j=1}^{q} d^{-1} N_f^{[l-1]}(r, D_j) + O(\log T_f(r) + \log r)$$

for every $r \in [0, \infty)$ excluding a set E' such that $\int_{E'} dt < \infty$.

Now consider the case where $R_0 < \infty$ and

$$\limsup_{r\to R_0} \frac{T_f(r)}{\log(1/(R_0 - r))} = \infty.$$

Since $T_f(r)$ is continuous and increasing, we may assume that $T_f(r) \ge 1$. By Lemma 4.1.7

$$T_f\left(r + \frac{R_0 - r}{eT_f(r)}\right) \le 2T_f(r)$$

outside a set E of r such that $\int_E 1/(R_0 - r)dr < \infty$. Taking $R = r + \frac{R_0 - r}{eT_f(r, r_0)}$, we obtain

$$(q - N - 1 - \epsilon)T_f(r) \leq \sum_{j=1}^{q} d^{-1} N_f^{[l-1]}(r, D_j) + K\left(\log^+ \frac{1}{R_0 - r} + \log^+ T_f(r)\right)$$

for every $r \in [0, R_0)$ excluding a set E such that $\int_E \frac{1}{R_0 - t} dt < \infty$. In both cases, we get

$$\sum_{j=1}^{q} \delta_{l-1}^{f,\star}(D_j) \leq N + 1.$$

Hence

$$\sum_{j=1}^{q} \delta_{l-1}^{f}(H_j) \leq \sum_{j=1}^{q} \delta_{l-1}^{f,\star}(H_j) \leq N + 1. \qquad \square$$

Note that in the second part of the theorem, we indeed proved the following second main theorem for mappings intersecting hypersurfaces.

Theorem. *Let $f : B(R_0) \subset \mathbb{C}^m \longrightarrow \mathbb{P}^n(\mathbb{C})$, $0 < R_0 \leq \infty$, be an algebraically nondegenerate meromorphic map, and let $D_1, \ldots, D_q$ be hypersurfaces of degrees $d_j, 1 \leq j \leq q$, in $\mathbb{P}^n(\mathbb{C})$ located in general position. Then, for all $\epsilon > 0$ and $0 < r < R_0$, we have*

$$(q - (n + 1 + \epsilon))T_f(r) \leq \sum_{j=1}^{q} d_j^{-1} N_f^{(l-1)}(r, D_j) + S(r),$$

where $l \leq 2^{n^2+4n} e^n d^{2n} (nI(\epsilon^{-1}))^n$, $d = l.c.m.\{d_1, \ldots, d_q\}$, and $S(r)$ is evaluated as follows:

(1) *In the case $R_0 < \infty$,*

$$S(r) \leq K\left(\log^+ \frac{1}{R_0 - r} + \log^+ T_f(r)\right)$$

for every $r \in [0, R_0)$ excluding a set E such that $\int_E \frac{1}{R_0 - t} dt < \infty$.

(2) *In the case $R_0 = \infty$,*

$$S(r) \leq K(\log^+ T_f(r) + \log r)$$

for every $r \in [0, \infty)$ excluding a set E' such that $\int_{E'} dt < \infty$.

Bibliography

[1] L. Ahlfors: Zur theorie der überlagerungsflachen, *Acta Math.*, 65 (1935), 157–194.
[2] L. Ahlfors: The theory of meromorphic curves, *Acta Soc. Sci. Fenn., Nova Ser. A*, 3(4) (1941), 171–183.
[3] J. L. Barbosa and A. G. Colares: Minimal surfaces in $\mathbb{R}^3$, Lecture Notes in Mathematics, 1195, Springer, 1986.
[4] S. Bernstein: Sur les surface défines au moyen de leur courbure moyenne ou totale, *Ann. Sci. Ec. Norm. Super.*, 27 (1910), 233–256.
[5] S. Bernstein: Sur les équations du calcul des variations, *Ann. Sci. Ec. Norm. Super.*, 29 (1912), 431–485.
[6] S. Bernstein: Sur un théorème de Géométrie et ses applications aux équations aux dérivées partielles du type elliptique, *Comm. Soc. Math. Kharkov (2 Sér.)*, 15 (1915–1917), 38–45.
[7] A. Biancofiore and W. Stoll: Another proof of the lemma of the logarithmic derivative in several complex variables. In: J. Fornaess (ed.) Recent developments in several complex variables, 29–45, 1981.
[8] R. Brody: Compact manifolds and hyperbolicity, *Trans. Am. Math. Soc.*, 235 (1978), 213–219.
[9] H. Cartan: Sur les systeme de fonctions holomorphes a variétés linéaires lacunaires et leurs applications, *Ann. Sci. Éc. Norm. Supér. (3)*, 45 (1928), 255–346.
[10] H. Cartan: Sur les zeros des combinaisons linéaires de p fonctions holomorphes données, *Mathematica (Cluj)*, 7 (1933), 5–31.
[11] W. Chen: Defect relations for degenerate meromorphic maps, *Trans. Am. Math. Soc.*, 319 (1990), 499–515.
[12] Y. Chen and E. Fried, Conformal deformations of a material surface. Preprint.
[13] X. Chen, Y. Li, Z. Liu and M. Ru: Curvature estimate on an open Riemann surface with the induced metric, *Math. Z.*, 298 (2021), 451–467.
[14] S. S. Chern: An elementary proof of the existence of isothermal parameters on a surface, *Proc. Am. Math. Soc.*, 6 (1955), 771–782.
[15] S. S. Chern: Complex analytic mappings of Riemann surfaces I, *Am. J. Math.*, 82 (1960), 323–337.
[16] S. S. Chern: Minimal surfaces in an Euclidean space of N-dimensions. In: Differential and Combinatorial Topology (A Symposium in Honor of Marston Morse), 187–198, Princeton Univ. Press, Princeton, N. J., 1965.
[17] S. S. Chern: On holomorphic mappings of Hermitian manifolds of the same dimension, *Proc. Symp. Pur. Math. Amer. Math. Soc.*, 11 (1968), 157–170.
[18] S. S. Chern: Simple proofs of two theorems on minimal surfaces, *Enseign. Math. (2)*, 15 (1969), 53–61.
[19] S. S. Chern: Holomorphic curves and minimal surfaces. In: Proc. Carolina Conf. on Holomorphic Mappings and Minimal Surfaces (Chapel Hill, N. C., 1970), 1–28, University of North Carolina, Department of Mathematics, Chapel Hill, N. C., 1970.
[20] S. S. Chern and R. Osserman: Complete minimal surfaces in euclidean n-space, *J. Anal. Math.*, 19 (1967), 15–34.
[21] M. Cowen and Ph. Griffiths: Holomorphic curves and metrics of nonnegative curvature, *J. Anal. Math.*, 29 (1976), 93–153.
[22] H. Federer: Geometric Measure Theory, Springer, New York, 1969.
[23] H. Fujimoto: The defect relations for the derived curves of a holomorphic curve in $\mathbb{P}^n(\mathbb{C})$, *Tohoku Math. J.*, 34 (1982), 141–160.
[24] H. Fujimoto: On the Gauss map of a complete minimal surface in $\mathbb{R}^m$, *J. Math. Soc. Jpn.*, 35(2) (1983), 279–288.
[25] H. Fujimoto: Value distribution of the Gauss maps of complete minimal surfaces in $\mathbb{R}^m$, *J. Math. Soc. Jpn.*, 35(4) (1983), 663–681.
[26] H. Fujimoto: Nonintegrated defect relation for meromorphic maps of complete Kähler manifolds into $\mathbb{P}^N(\mathbb{C}) \times \cdots \times \mathbb{P}^N(\mathbb{C})$, *Jpn. J. Math. New Ser.*, 11(2) (1985), 233–264.

https://doi.org/10.1515/9783110989557-007

[27] H. Fujimoto: On the number of exceptional values of the Gauss maps of minimal surfaces, *J. Math. Soc. Jpn.*, 40(2) (1988), 235–247.

[28] H. Fujimoto: Examples of complete minimal surfaces in R^m whose Gauss maps omit $m(m+1)/2$ hyperplanes in general position, *Sci. Rep. Kanazawa Univ.*, 33(2) (1988), 37–43.

[29] H. Fujimoto: On the value distribution of the Gauss map of minimal surfaces in $\mathbb{R}^n$ (Japanese), *Sūgaku*, 40(4) (1988), 312–321.

[30] H. Fujimoto: Modified defect relations for the Gauss map of minimal surfaces, *J. Differ. Geom.*, 29(2) (1989), 245–262.

[31] H. Fujimoto: Modified defect relations for the Gauss map of minimal surfaces II, *J. Differ. Geom.*, 31(2) (1990), 365–385.

[32] H. Fujimoto: Modified defect relations for the Gauss map of minimal surface III, *Nagoya Math. J.*, 124 (1991), 13–40.

[33] H. Fujimoto: On the Gauss curvature of minimal surfaces, *J. Math. Soc. Jpn.*, 44(3) (1992), 427–439.

[34] H. Fujimoto: Value Distribution Theory on the Gauss Map of Minimal Surfaces in $\mathbb{R}^m$, Aspect of Mathematics, E21, Vieweg, 1993.

[35] M. Gaffney: A special Stokes theorem for complete Riemannian manifolds, *Ann. Math.*, 60 (1954), 140–145.

[36] M. Green: Some Picard theorems for holomorphic maps to algebraic varieties, *Am. J. Math.*, 97 (1975), 43–75.

[37] M. Green: The hyperbolicity of the complement of $2n+1$ hyperplanes in general position in $\mathbb{P}^n$ and related results, *Proc. Am. Math. Soc.*, 66 (1977), 109–113.

[38] R. E. Greene and H. Wu: Function theory on manifolds which possess a pole, Lecture Notes in Mathematics, 699, Springer, 1979.

[39] Ph. Griffiths and J. Harris: Principle of algebraic geometry, Wiley, 1994.

[40] W. Hayman: Meromorphic Functions, Oxford University Press, 1964.

[41] E. Heinz: Über die Lösungen der Minimalflächengleichung, Nachr. Akad. Wiss. Gött., 3.Iath.-Ph.vs. Ki. (1952), 51–56.

[42] E. Hopf: On an inequality for minimal surfaces $z = z(x,y)$, *J. Ration. Mech. Anal.*, 2 (1953), 519–522.

[43] L. Jin and M. Ru: Values of Gauss maps of complete minimal surfaces in $\mathbb{R}^m$ on annular ends, *Trans. Am. Math. Soc.*, 359 (2007), 1527–1546.

[44] L. Jin and M. Ru: Algebraic curves and the Gauss map of algebraic minimal surfaces, *Differ. Geom. Appl.*, 25 (2007), 701–712.

[45] L. Karp: Subharmonic functions on real and complex manifolds, *Math. Z.*, 179 (1982), 535–554.

[46] Y. Kawakami: On the maximal number of exceptional values of Gauss maps for various classes of surfaces, *Math. Z.*, 274 (2013), 1249–1260.

[47] S. Lang: Introduction to complex hyperbolic spaces, Springer, 1987.

[48] X. Mo: Value distribution of the Gauss map and the total curvature of complete minimal surface in $\mathbb{R}^m$, *Pac. J. Math.*, 163(1) (1994), 159–174.

[49] X. Mo and R. Osserman: On the Gauss map and total curvature of complete minimal surfaces and an extension of Fujimoto's theorem, *J. Differ. Geom.*, 31(2) (1990), 343–355.

[50] R. Nevanlinna: Zur Theorie der meromorphen Funktionen, *Acta Math.*, 46 (1925), 1–99.

[51] R. Nevanlinna: Le théorème de Picard-Borel et la théorie des fonctions meromorphes, Gauthiers-Villars, Paris, 1929.

[52] R. Nevanlinna: Analytic Functions, Springer, 1970.

[53] E. I. Nochka: On a theorem from linear algebra, Izv. Akad. Nauk. Modav. SSR Ser. Fiz. Teklam Mat. Nauk (1982), 29–33.

[54] E. I. Nochka: On the theory of meromorphic curves, *Dokl. Akad. Nauk SSSR*, 269(3) (1983), 547–552.

[55] R. Osserman: An analogue of the Heinz-Hopf inequality, *J. Math. Mech.*, 8 (1959), 383–385.

[56] R. Osserman: Proof of a conjecture of Nirenberg, *Commun. Pure Appl. Math.*, 12 (1959), 229–232.

[57] R. Osserman: Remarks on minimal surfaces, *Commun. Pure Appl. Math.*, 12 (1959), 233–239.

[58] R. Osserman: On the Gauss curvature of minimal surfaces, *Trans. Am. Math. Soc.*, 96 (1960), 115–128.
[59] R. Osserman: Minimal surfaces in the large, *Comment. Math. Helv.*, 35 (1961), 65–76.
[60] R. Osserman: On complete minimal surfaces, *Arch. Ration. Mech. Anal.*, 13 (1963), 392–404.
[61] R. Osserman: Global properties of minimal surfaces in E3 and En, *Ann. Math (2)*, 80 (1964), 340–364.
[62] R. Osserman: Minimal surfaces, Gauss maps, total curvature, eigenvalue estimates, and stability. In: The Chern Symposium 1979 (Proc. Internat. Sympos., Berkeley, Calif., 1979), 199–227, Springer, New York-Berlin, 1980.
[63] R. Osserman: A survey of minimal surfaces, 2nd ed., Dover Publications, Inc., New York, 1986. ISBN 0-486-64998-9, vi+207 pp.
[64] R. Osserman: A new variant of the Schwarz-Pick-Ahlfors lemma, *Manuscr. Math.*, 100(2) (1999), 123–129.
[65] R. Osserman: A sharp Schwarz inequality on the boundary, *Proc. Am. Math. Soc.*, 128(12) (2000), 3513–3517.
[66] R. Osserman and M. Ru: An estimate for the Gauss curvature of minimal surfaces in $\mathbb{R}^m$ whose Gauss map omits a set of hyperplanes, *J. Differ. Geom.*, 46(3) (1997), 578–593.
[67] M. Ru: On the Gauss map of minimal surfaces immersed in $\mathbb{R}^n$, *J. Differ. Geom.*, 34(2) (1991), 411–423.
[68] M. Ru: On the Gauss map of minimal surfaces with finite total curvature, *Bull. Aust. Math. Soc.*, 44 (1991), 225–232.
[69] M. Ru: Integral points and the hyperbolicity of the complement of hypersurfaces, *J. Reine Angew. Math.*, 442 (1993), 163–176.
[70] M. Ru: Geometric and arithmetic aspects of $\mathbb{P}^n$ minus hyperplanes, *Am. J. Math.*, 117 (1995), 307–321.
[71] M. Ru: On a general form of the second main theorem, *Trans. Am. Math. Soc.*, 349 (1997), 5093–5105.
[72] M. Ru: A defect relation for holomorphic curves intersecting hypersurfaces, *Am. J. Math.*, 126(1) (2004), 215–226.
[73] M. Ru: Holomorphic curves into algebraic varieties, *Ann. Math. (2)*, 169(1) (2009), 255–267.
[74] M. Ru: A defect relation for holomorphic curves intersecting general divisors in projective varieties, *J. Geom. Anal.*, 26(4) (2016), 2751–2776.
[75] M. Ru: Recent progress in Nevanlinna theory, *J. Jiangxi Norm. Univ. Nat. Sci.*, 42(1) (2018), 1–11.
[76] M. Ru: Nevanlinna theory and its relation to Diophantine approximation, 2nd ed., World Scientific, 2021.
[77] M. Ru and S. Sogome: Non-integrated defect relation for meromorphic maps of complete Kähler manifold intersecting hypersurfaces in $\mathbb{P}^n(\mathbb{C})$, *Trans. Am. Math. Soc.*, 364(3) (2012), 1145–1162.
[78] M. Ru and P. Vojta: A birational Nevanlinna constant and its consequences, *Am. J. Math.*, 142 (2020), 957–991.
[79] B. Seguin and E. Fried: Coordinate-free compatibility conditions for deformations of material surfaces, *Mech. Mater.*, 166, 104193 (2022).
[80] B. V. Shabat: Distribution of values of holomorphic mappings, Translations of Mathematical Monographs, 61, American Mathematical Society, 1985.
[81] B. Shiffman: Nevanlinna defect relations for singular divisors, *Invent. Math.*, 31 (1975), 155–182.
[82] B. Shiffman: Holomorphic curves in algebraic manifolds, *Bull. Am. Math. Soc.*, 83 (1977), 553–568.
[83] W. Stoll, in Introduction to value distribution theory of meromorphic maps, Lecture notes in Math., 950, 210–359, 1982.
[84] T. V. Tan and V. V. Truong: A non-integrated defect relation for meromorphic mappings from complete Kähler manifolds into a projective variety intersecting hypersurfaces, *Bull. Sci. Math.*, 136 (2012), 111–126.
[85] M. Troyanov: The Schwarz lemma for nonpositively curved Riemannian surfaces, *Manuscr. Math.*, 72 (1991), 251–256.
[86] A. Vitter: The lemma of the logarithmic derivative in several complex variables, *Duke Math. J.*, 44 (1977), 89–104.
[87] A. Weitsman and F. Xavier: Some function theoretic properties of the Gauss map for hyperbolic complete minimal surfaces, *Mich. Math. J.*, 34(2) (1987), 275–283.

[88] F. Xavier: The Gauss map of a complete nonflat minimal surface cannot omit 7 points of the sphere, *Ann. Math. (2)*, 113(1) (1981), 211–214.

[89] F. Xavier: Erratum to the paper: "The Gauss map of a complete nonflat minimal surface cannot omit 7 points of the sphere", *Ann. Math. (2)*, 113(1) (1981), 211–214, [Ann. of Math. (2) 113(1) (1981), 211--214, MR 82b:53015].

[90] S. T. Yau: Some function-theoretic properties of complete Riemannian manifold and their applications to geometry, *Indiana Univ. Math. J.*, 25(7) (1976), 659–670.

[91] S. T. Yau: A general Schwarz lemma for Kähler manifolds, *Am. J. Math.*, 100 (1978), 197–203.

Index

https://doi.org/10.1515/9783110989557-008

De Gruyter Studies in Mathematics

Volume 46/2
Steven Lord, Edward McDonald, Fedor Sukochev, Dmitriy Zanin
Singular Traces. Volume 2: Trace Formulas, 2nd Edition, 2023
ISBN 978-3-11-070000-8, e-ISBN 978-3-11-070017-6, e-ISBN (ePUB) 978-3-11-070024-4

Volume 91
Johnny Henderson, Rodica Luca
Boundary Value Problems for Second-Order Finite Difference Equations and Systems, 2023
ISBN 978-3-11-103931-2, e-ISBN 978-3-11-104037-0, e-ISBN (ePUB) 978-3-11-104045-5

Volume 91
Johnny Henderson, Rodica Luca
Boundary Value Problems for Second-Order Finite Difference Equations and Systems, 2023
ISBN 978-3-11-103931-2, e-ISBN 978-3-11-104037-0, e-ISBN (ePUB) 978-3-11-104045-5

Volume 90
William M. McGovern
Representation Theory and Geometry of the Flag Variety, 2022
ISBN 978-3-11-076690-5, e-ISBN 978-3-11-076694-3, e-ISBN (ePUB) 978-3-11-076696-7

Volume 89
Marco Abate
Holomorphic Dynamics on Hyperbolic Riemann Surfaces, 2022
ISBN 978-3-11-060105-3, e-ISBN 978-3-11-060197-8, e-ISBN (ePUB) 978-3-11-059874-2

Volume 88
Joel L. Schiff
Topics in Complex Analysis, 2022
ISBN 978-3-11-075769-9, e-ISBN 978-3-11-075782-8, e-ISBN (ePUB) 978-3-11-075787-3

Volume 66
Igor V. Nikolaev
Noncommutative Geometry. A Functorial Approach, 2nd Edition, 2022
ISBN 978-3-11-078860-0, e-ISBN 978-3-11-078870-9, e-ISBN (ePUB) 978-3-11-078881-5

Volume 87
René Erlin Castillo
Fourier Meets Hilbert and Riesz. An Introduction to the Corresponding Transforms, 2022
ISBN 978-3-11-078405-3, e-ISBN 978-3-11-078409-1, e-ISBN (ePUB) 978-3-11-078412-1

www.degruyter.com

www.ingramcontent.com/pod-product-compliance
Ingram Content Group UK Ltd.
Pitfield, Milton Keynes, MK11 3LW, UK
UKHW050953290926
15470UKWH00006B/1797

9 783110 999822